Lecture Notes in Computer Science

Commenced Publication in 1973
Founding and Former Series Editors:
Gerhard Goos, Juris Hartmanis, and Jan van

Lecture Notes in Computer Science 7401

Jin-Kao Hao   Pierrick Legrand
Pierre Collet   Nicolas Monmarché
Evelyne Lutton   Marc Schoenauer (Eds.)

Artificial Evolution

10th International Conference
Evolution Artificielle, EA 2011
Angers, France, October 24-26, 2011
Revised Selected Papers

 Springer

Volume Editors

Jin-Kao Hao
Université d'Angers, France
E-mail: jin-kao.hao@univ-angers.fr

Pierrick Legrand
Université Bordeaux Segalen, France
E-mail: pierrick.legrand@u-bordeaux2.fr

Pierre Collet
Université de Strasbourg, France
E-mail: pierre.collet@unistra.fr

Nicolas Monmarché
Ecole Polytechnique de l'Université de Tours, France
E-mail: nicolas.monmarche@univ-tours.fr

Evelyne Lutton
Marc Schoenauer
INRIA Saclay - Île-de-France, Orsay Cedex, France
E-mail: {evelyne.lutton, marc.schoenauer}@inria.fr

ISSN 0302-9743   e-ISSN 1611-3349
ISBN 978-3-642-35532-5   e-ISBN 978-3-642-35533-2
DOI 10.1007/978-3-642-35533-2
Springer Heidelberg Dordrecht London New York

Library of Congress Control Number: 2012953904

CR Subject Classification (1998): I.2.8-9, F.1.1-2, F.2.1-2, G.1.6-7, I.5.1, H.3.3-4

LNCS Sublibrary: SL 1 – Theoretical Computer Science and General Issues

Typesetting: Camera-ready by author, data conversion by Scientific Publishing Services, Chennai, India

Printed on acid-free paper

Springer is part of Springer Science+Business Media (www.springer.com)

Preface

This LNCS volume includes the best papers presented at the 10th Biennial International Conference on Artificial Evolution, EA[1] 2011, held in Angers (France). Previous EA editions took place in Strasbourg (2009), Tours (2007), Lille (2005), Marseille (2003), Le Creusot (2001), Dunkerque (1999), Nimes (1997), Brest (1995), and Toulouse (1994).

Authors had been invited to present original work relevant to artificial evolution, including, but not limited to: evolutionary computation, evolutionary optimization, co-evolution, artificial life, population dynamics, theory, algorithmics and modeling, implementations, application of evolutionary paradigms to the real world (industry, biosciences), other biologically inspired paradigms (swarm, artificial ants, artificial immune systems, cultural algorithms), memetic algorithms, multi-objective optimization, constraint handling, parallel algorithms, dynamic optimization, machine learning and hybridization with other soft computing techniques.

Each submitted paper was reviewed by three members of the International Program Committee. Among the 64 submissions received, 33 papers were selected for oral presentation and 10 other papers for poster presentation. As for the previous editions (see LNCS volumes 1063, 1363, 1829, 2310, 2936, 3871, 4926 and 5975), a selection of the best papers (19 papers, hence an acceptance rate of less than 30%) which were presented at the conference and further revised is published in this volume of Springer's LNCS series.

To celebrate the tenth anniversary of EA, we are grateful to two of the cofounders of the EA series, Evelyne Lutton and Marc Schoenauer both from INRIA (France), who accepted to give a talk on "Twenty Years of Artificial Evolution in France" (Jean-Marc Alliot from ENAC Toulouse is the other cofounder). We would also like to express our sincere gratitude to our invited speaker René Doursat from Institut des Systèmes Complexes – Paris, who gave the talk "Artificial Evo-Devo: Bringing Back Self-Organized Multi-Agent Systems into Evolutionary Computation."

The success of the conference resulted from the input of many people to whom we would like to express our appreciation: the members of Program Committee for their careful reviews that ensure the quality of the selected papers and the conference; the members of the Organizing Committee for their efficient work and dedication assisted by Catherine Pawlonski-Boisseau and Christine Bardaine and others from the Computer Science Department of the University of Angers; the members of the Steering Committee for their valuable assistance; Sylvie Reverdy from Centre de Congrès d'Angers for her very efficient actions; and Marc Schoenauer for his support with the MyReview system.

---

[1] As for previous editions of the conference, the EA acronym is based on the original French name "Évolution Artificielle."

We take this opportunity to thank the different partners whose financial and material support contributed to the organization of the conference: Université d'Angers, Conseil Général "Maine et Loire", Ville d'Angers, Angers Loire Métropole, Région "Pays de La Loire", INRIA, Ministère de l'Enseignement Supérieur et de la Recherche, Laboratoire LERIA, Centre de Congrès d'Angers.

Last but not least, we thank all the authors who submitted their research papers to the conference, the authors of accepted papers who attended the conference to present their work, and the attendees. Thank you all.

July 2012

Jin-Kao Hao
Pierrick Legrand
Pierre Collet
Nicolas Monmarché
Evelyne Lutton
Marc Schoenauer

Photo by Nicolas Monmarché.

Évolution Artificielle 2011 – EA 2011

October 24–26, 2011
Angers, France
10th International Conference on Artificial Evolution

Steering Committee

Pierre Collet	Université de Strasbourg
Evelyne Lutton	INRIA Saclay—Île-de-France
Nicolas Monmarché	Université François Rabelais de Tours
Marc Schoenauer	INRIA Saclay—Île-de-France

Organizing Committee

Anne Auger (Submissions)
Béatrice Duval (Sponsors)
Jin-Kao Hao (General Chair)
Laetitia Jourdan (Publicity)
Pierrick Legrand (Proceedings / LNCS publication)
Jean-Michel Richer (Webmaster)
Sébastien Vérel (Treasurer)

Jean-Michel Richer and Jin-Kao Hao.
Photo by Nicolas Monmarché.

International Program Committee

Invited Talks

1 - **René Doursat**, Institut des Systèmes Complexes - Paris (France)

Artificial Evo-Devo: Bringing Back Self-Organized Multi-Agent Systems into Evolutionary Computation

 Multicellular organisms and social insect constructions are rather unique examples of naturally evolved systems that exhibit both self-organization and a strong architecture. Can we export their precise self-formation capabilities to technological systems ? I have proposed a new research field called "morphogenetic engineering" [1], which explores the artificial design and implementation of complex, heterogeneous morphologies capable of developing without central planning or external lead. Particular emphasis is set on the programmability and controllability of self-organization, properties that are often underappreciated in complex systems science, while, conversely, the benefits of multi-agent self-organization are often underappreciated in engineering methodologies, including evolutionary computation. In this talk I presented various examples of morphogenetic engineering, in particular multi-agent systems inspired by biological development based on gene regulation networks, and self-construction of graph topologies based on "programmed attachment" rules. Potential applications range from robotic swarms to autonomic networks and socio-technical systems. In all cases, the challenge is to "meta-design," especially through an evolutionary search, the proper set of rules followed by each agent of a complex system on how to interact with the other agents and the environment. Whether "offline" (slow time scale), where agents always share the same genotype, or "online" (fast time scale), where agent types may diverge and specialize dynamically, it constitutes an inherently massively parallel "artificial evo-devo" (evolutionary developmental) problem.

[1]. Doursat, R., Sayama, H. and Michel, O., eds. (2011) Morphogenetic Engineering: Toward Programmable Complex Systems, in NECSI "Studies on Complexity" Series, Springer-Verlag, in press.

2 - Evelyne Lutton and Marc Schoenauer, INRIA (France)

Twenty Years of Artificial Evolution in France

Sponsoring Institutions

- Université d'Angers
- Conseil Général "Maine et Loire"
- Ville d'Angers
- Angers Loire Métropole
- Région "Pays de La Loire"
- INRIA
- Ministère de l'Enseignement Supérieur et de la Recherche
- Association Evolution Artificielle http://www.lifl.fr/EA/

Table of Contents

Combinatorial Optimization

Learning and Parameter Tuning

New Nature Inspired Models

Probabilistic Algorithms

Theory and Evolutionary Search

Applications

An Immigrants Scheme Based on Environmental Information for Ant Colony Optimization for the Dynamic Travelling Salesman Problem

Michalis Mavrovouniotis[1] and Shengxiang Yang[2]

[1] Department of Computer Science, University of Leicester
University Road, Leicester LE1 7RH, United Kingdom
mm251@mcs.le.ac.uk
[2] Department of Information Systems and Computing, Brunel University
Uxbridge, Middlesex UB8 3PH, United Kingdom
shengxiang.yang@brunel.ac.uk

Abstract. Ant colony optimization (ACO) algorithms have proved to be powerful methods to address dynamic optimization problems. However, once the population converges to a solution and a dynamic change occurs, it is difficult for the population to adapt to the new environment since high levels of pheromone will be generated to a single trail and force the ants to follow it even after a dynamic change. A good solution is to maintain the diversity via transferring knowledge to the pheromone trails. Hence, we propose an immigrants scheme based on environmental information for ACO to address the dynamic travelling salesman problem (DTSP) with traffic factor. The immigrants are generated using a probabilistic distribution based on the frequency of cities, constructed from a number of ants of the previous iteration, and replace the worst ants in the current population. Experimental results based on different DTSP test cases show that the proposed immigrants scheme enhances the performance of ACO by the knowledge transferred from the previous environment and the generation of guided diversity.

1 Introduction

Ant colony optimization (ACO) algorithms have proved to be powerful meta-heuristics for solving difficult real-world optimization problems under static environments [2,12]. ACO is inspired from real ant colonies, where a population of ants searches for food from their nest. Ants communicate via their pheromone trails and they are able to track shortest paths between their nest and food sources. Inspired from this behaviour, ACO algorithms are able to locate the optimum, or a near optimum solution, in stationary optimization problems, efficiently [10]. However, in many real-world problems we have to deal with dynamic environments, where the optimum is moving, and the objective is not only to locate the optimum but to track it also [7]. ACO algorithms face a serious challenge in dynamic optimization problems (DOPs) because they lose their adaptation capabilities once the population has converged [1].

J.-K. Hao et al. (Eds.): EA 2011, LNCS 7401, pp. 1–12, 2012.

Over the years, several approaches have been developed for ACO algorithms to enhance their performance for DOPs, such as local and global restart strategies [6], memory-based approaches [5], pheromone update schemes to maintain diversity [3], and immigrants schemes to increase diversity [8,9]. Most of these approaches have been applied in different variations of the dynamic travelling salesman problem (DTSP), since it has many similarities with many real-world applications [11]. Among these approaches, immigrants schemes have been found beneficial to deal with DTSPs, where immigrant ants are generated and used to replace other ants of the current population [8,9]. The most important concern when applying immigrants schemes is how to generate immigrants.

In this paper, we propose an immigrants scheme which is based on environmental information for ACO for the DTSP. The environmental information-based immigrants ACO (EIIACO) selects the first n best ants from the previous environment and creates a probabilistic distribution based on the frequency of cities that appear next to each other. Using the information obtained from the previous population, immigrants are generated to replace the worst ants in the current population. The introduced immigrants transfer knowledge and can guide the population toward the promising regions after a change occurs. It is expected that the environmental information-based immigrants scheme may enhance the performance of ACO for DTSPs, especially when the environments before and after a change are similar.

The remaining of this paper is organized as follows. Section 2 describes the DTSP used in the experimental study in this paper. Section 3 describes existing ACO algorithms for the DTSP, which are also investigated as peer algorithms in the experiments. Section 4 describes the proposed EIIACO algorithm. The experimental results and analysis are presented in Section 5. Finally, the conclusions and relevant future work are presented in Section 6.

2   DTSP with the Traffic Jam

The TSP is the most fundamental and well-known NP-hard combinatorial optimization problem. It can be described as follows: Given a collection of cities, we need to find the shortest path that starts from one city and visits each of the other cities once and only once before returning to the starting city.

In this paper, we generate a DTSP via introducing the traffic factor. We assume that the cost of the link between cities i and j is $C_{ij} = D_{ij} \times F_{ij}$, where D_{ij} is the normal travelled distance and F_{ij} is the traffic factor between cities i and j. Every f iterations a random number R in $[F_L, F_U]$ is generated probabilistically to represent traffic between cities, where F_L and F_U are the lower and upper bounds of the traffic factor, respectively. Each link has a probability m to add traffic by generating a different R each time, such that $F_{ij} = 1 + R$, where the traffic factor of the remaining links is set to 1 (indicates no traffic). Note that f and m denote the frequency and magnitude of the changes in the dynamic environment, respectively. The TSP becomes more challenging and realistic when it is subject to a dynamic environment. For example, a traffic factor closer to the upper bound F_U represents rush hour periods which increases the travelled distance significantly. On the other hand, a traffic factor closer to the lower bound F_L represents normal hour periods which increases the travelled distance slightly.

3 ACO for the DTSP

3.1 Standard ACO

The standard ACO (S-ACO) algorithm, i.e., Max-Min AS (MMAS), consists of a population of μ ants [12]. Initially, all ants are placed to a randomly selected city and all pheromone trails are initialized with an equal amount of pheromone. With a probability $1 - q_0$, where $0 \leq q_0 \leq 1$ is a parameter of the decision rule, ant k chooses the next city j, while its current city is i, probabilistically, as follows:

$$p_{ij}^k = \frac{[\tau_{ij}]^\alpha [\eta_{ij}]^\beta}{\sum_{l \in N_i^k} [\tau_{il}]^\alpha [\eta_{il}]^\beta}, \text{if } j \in N_i^k, \tag{1}$$

where τ_{ij} and $\eta_{ij} = 1/C_{ij}$ are the existing pheromone trails and heuristic information available a priori between cities i and j, respectively, N_i^k denotes the neighbourhood of cities of ant k that have not yet been visited when its current city is i, and α and β are the two parameters that determine the relative influence of pheromone trail and heuristic information, respectively. With the probability q_0, ant k chooses the next city with the maximum probability, i.e., $[\tau]^\alpha [\eta]^\beta$, and not probabilistically as in Eq. (1).

Later on, the best ant retraces the solution and deposits pheromone according to its solution quality on the corresponding trails as follows:

$$\tau_{ij} \leftarrow \tau_{ij} + \Delta\tau_{ij}^{best}, \forall (i,j) \in T^{best}, \tag{2}$$

where $\Delta\tau_{ij}^{best} = 1/C^{best}$ is the amount of pheromone that the best ant deposits and C^{best} is the tour cost of T^{best}. However, before adding any pheromone, a constant amount of pheromone is deduced from all trails due to the pheromone evaporation such that, $\tau_{ij} \leftarrow (1 - \rho)\tau_{ij}, \forall (i,j)$, where $0 < \rho \leq 1$ is the rate of evaporation. The pheromone trail values are kept to the interval $[\tau_{min}, \tau_{max}]$ and they are re-initialized to τ_{max} every time the algorithm shows a stagnation behaviour, where all ants follow the same path, or when no improved tour has been found for several iterations [12].

The S-ACO algorithm faces a serious challenge when it is applied to DTSPs. The pheromone trails of the previous environment will not make sense to the new one. From the initial iterations the population of ants will eventually converge to a solution, and, thus, high intensity of pheromone trails will be generated into a single path. The pheromone trails will influence ants to the current path even after a dynamic change. The pheromone evaporation is the only mechanism used to eliminate the pheromone trails generated previously, and help ants to adapt to the new environment. Therefore, S-ACO requires a sufficient amount of time in order to recover when a dynamic change occurs.

3.2   Population-Based ACO

The population-based ACO (P-ACO) algorithm is the memory based version of ACO [5]. It differs from the S-ACO algorithm described above since it follows a different framework. The algorithm maintains a population of ants (solutions), called population-list (memory), which is used to update pheromone trails without any evaporation.

The initial phase and the first iterations of the P-ACO algorithm work in the same way as with the S-ACO algorithm. The pheromone trails are initialized with an equal amount of pheromone and the population-list of size K is empty. For the first K iterations, the iteration-best ant deposits a constant amount of pheromone, using Eq. (2) with $\Delta\tau_{ij}^{best} = (\tau_{max} - \tau_{init})/K$. Here, τ_{max} and τ_{init} denote the maximum and initial pheromone amount, respectively. This positive update procedure is performed whenever an ant enters the population-list. On iteration $K + 1$, the ant that has entered the population-list first, i.e., the oldest ant, needs to be removed in order to make room for the new one, and its pheromone trails are reduced by $\Delta\tau_{ij}^{best}$, which equals to the amount added when it entered the population-list before.

The population-list is a long-term memory, denoted as k_{long}, since it may contain ants from previous environments that survive in more than one iteration. Therefore, when a change occurs, the ants stored in the population-list are re-evaluated in order to be consistent with the new environment where different traffic factors F_{ij} are introduced. The pheromone trails are updated accordingly using the ants currently stored in the population-list.

The P-ACO algorithm has a more aggressive mechanism to eliminate previously generated trails since the corresponding pheromone of the ants that are replaced by other ants in the population-list are removed directly. However, they face the same challenge with S-ACO because identical ants may be stored in the population-list and dominate the search space with a high intensity of pheromone to a single trail. However, the P-ACO algorithm may be beneficial when a new environment is similar to an old one because the solutions stored in the population-list from the previous environments will guide ants towards promising areas in the search space.

4   Immigrants Based on Environmental Information

Many immigrants schemes have been found beneficial in genetic algorithms (GAs) for binary-encoded DOPs [4,13,14]. Therefore, to handle the problems described in S-ACO and P-ACO algorithms when addressing DTSPs, immigrants schemes are integrated with ACO since they maintain a certain level of diversity during the execution and transfer knowledge from previous environments [8,9]. The immigrants are integrated within P-ACO as follows. A short-term memory is used, denoted as k_{short}, where the ants survive only for one iteration. All the ants of the current iteration replace the old ones, instead of only replacing the oldest one as in P-ACO. Then, a predefined portion of the worst ants are replaced by immigrant ants in k_{short}. When ants are removed from k_{short}, a

---

Algorithm 1. Generate Frequency Of Cities

1: **input** k_{short} // short term memory
2: **input** n // size of k_{short}
3: **for** $k = 1$ to n **do**
4:    **for** $i = 1$ to l **do**
5:       $city\_one\_id = ant[k].tour[i]$
6:       $city\_two\_id = ant[k].tour[i + 1]$
7:       $frequency\_of\_cities[city\_one\_id][city\_two\_id]+=1$
8:       $frequency\_of\_cities[city\_two\_id][city\_one\_id]+=1$
9:    **end for**
10: **end for**
11: **return** $frequency\_of\_cities$

---

negative update is made to their pheromone trails and when new ants are added to k_{short}, a positive update is made to their pheromone trails as in P-ACO.

Different immigrants schemes were integrated with P-ACO, such as the traditional immigrants [8], where immigrant ants are generated randomly, the elitism-based immigrants [8], where immigrant ants are generate using the best ant from k_{short}, and the memory-based immigrants [9], where immigrant ants are generated using the best ant from k_{long}. The information obtained from the elitism- and memory-based immigrants to transfer knowledge is based on individual information (one ant). The proposed EIIACO algorithm generates immigrants using environmental information (population of ants) to transfer knowledge from the previous environment to a new one. EIIACO follows the same framework with other ACO algorithms based on immigrants schemes, as described above, but differs in the way immigrant ants are generated.

Environmental information-based immigrants are generated using all the ants stored in k_{short} of the previous environment. Within EIIACO, a probabilistic distribution based on the frequency of cities is extracted, representing information of the previous environment, which is used as the base to generate immigrant ants. The frequency vector of each city c_i, i.e, $\boldsymbol{D}_{c_i}$, is constructed by taking the ants of k_{short} as a dataset and locating city c_i from them. The successor and predecessor cities, i.e., c_{i-1} and c_{i+1}, respectively, of city c_i are obtained and update $\boldsymbol{D}_{c_i}$ accordingly. For example, one is added to the corresponding position $i - 1$ and $i + 1$ in $\boldsymbol{D}_{c_i}$. The process is repeated for all cities and a table $S = (\boldsymbol{D}_{c_1}, \ldots, \boldsymbol{D}_{c_l})$ is generated (where l is the number of cities) as represented in Algorithm 1.

An environmental information-based immigrant ant, i.e., $A_{eii} = (c_1, \ldots, c_l)$, is generated as follows. First, randomly select the start city c_1; then, the probabilistic distribution of $\boldsymbol{D}_{c_{i-1}} = (d_1, \ldots, d_l)$ is used to select the next city c_i probabilistically as follows:

$$p_i = \frac{d_i}{\sum_{j \in \boldsymbol{D}_{c_{i-1}}} d_j}, \text{if } i \in \boldsymbol{D}_{c_{i-1}}, \tag{3}$$

Algorithm 2. Generate Environmental Information-Based Immigrant

```
1:  input l                // number of cities
2:  input k                // immigrant ant identifier
3:  input frequency_of_cities    // see Algorithm 1
4:  step = 1               // counter for construction step
5:  ant[k].tour[step] = random[1, l]
6:  while step < l do
7:      step+=1
8:      current_city = ant[k].tour[step − 1]
9:      sum_probabilities = 0.0
10:     for j = 1 to l do
11:         if ant[k].visited[j] then
12:             probability[j] = 0.0
13:         else
14:             probability[j] = frequency[current_city][j]
15:             sum_probabilities += probability[j]
16:         end if
17:     end for
18:     if sum_probabilities = 0.0 then
19:         selected = random[1, l]
20:         while ant[k].visited[selected] do
21:             selected = random[1, l]
22:         end while
23:         ant[k].tour[step] = selected
24:     else
25:         r = random[0, sum_probabilities]
26:         selected = 1
27:         p = probability[selected]
28:         while p < r do
29:             selected+=1
30:             p += probability[selected]
31:         end while
32:         ant[k].tour[step] = selected
33:     end if
34: end while
35: ant[k].tour[l + 1] = ant[k].tour[1]
36: return ant[k]          // generated immigrant ant
```

where d_i is the frequency number where city c_i appears before or after city c_{i-1}. Note that all cities currently selected and stored in A_{eii} have a probability of 0.0 to be selected since they are already visited. In the case where the sum of $p_i = 0.0$, which means that all cities in $\boldsymbol{D}_{c_{i-1}}$ are visited, a random city j that has not been visited yet is selected. This process is repeated until all cities are used in order to generate a valid immigrant ant based on the environmental information, as represented in Algorithm 2. During lines 7–17, the probabilistic distribution is generated, during lines 18–24, the next city is selected randomly from the unvisited cities, and during lines 25–33, the next city is selected probabilistically from the frequency of cities generated from Algorithm 1. Note that in line 35 the first city stored in the ant is added to the end since the TSP tour is cyclic.

5   Simulation Experiments

5.1   Experimental Setup

In the experiments, we compare the proposed EIIACO algorithm with the S-ACO and P-ACO algorithms, which are described in Section 3. Our implementation follows the guidelines of the ACOTSP[1] framework. All the algorithms have been applied to the ei176, kroA100, and kroA200 problem instances, obtained from TSPLIB[2]. Most of the parameters have been optimized and obtained from our preliminary experiments while others have been inspired from the literature [5,8,9]. For all algorithms, $\mu = 25$ ants are used, $\alpha = 1$, $\beta = 5$ and $q_0 = 0.0$ (except P-ACO where $q_0 = 0.9$). Moreover, for S-ACO, $\rho = 0.2$. For P-ACO, $\tau_{max} = 1.0$, and the size of k_{long} is 3. For EIIACO, the size of k_{short} is 10 and four immigrant ants are generated. For each algorithm on a DTSP instance, $N = 30$ independent runs were executed on the same dynamic changes. The algorithms were executed for $G = 1000$ iterations and the overall offline performance of an algorithm is calculated as follows:

$$P_{offline} = \frac{1}{G} \sum_{i=1}^{G} \left(\frac{1}{N} \sum_{j=1}^{N} P_{ij}^* \right), \tag{4}$$

where P_{ij}^* defines the tour cost of the best ant since the last dynamic change of iteration i of run j [7].

The value of f was set to 20 and 100, which indicates fast and slowly changing environments, respectively. The probability of m was set to 0.1, 0.25, 0.5, and 0.75, which indicates the degree of environmental changes from small, to medium, to large, respectively. The intervals of the traffic factor were set to $F_L = 0$ and $F_U = 5$. As a result, 8 dynamic environments, i.e., 2 values of $f \times$ 4 values of m, were generated from each stationary TSP instance, to systematically analyze the adaptation and searching capability of each algorithm in the DTSP.

5.2   Experimental Results and Analysis

The experimental results regarding the overall offline performance of the algorithms for DTSPs are presented in Table 1. The corresponding statistical results of two-tailed t-test with 58 degree of freedom at a 0.05 level of significance are presented in Table 2, where "s+" or "s−" means that the first or the second algorithm is significantly better, respectively, and "+" or "−" means that the first or the second algorithm is insignificantly better, respectively. Moreover, to

---

[1] http://www.aco-metaheuristic.org/aco-code/
[2] http://comopt.ifi.uni-heidelberg.de/software/TSPLIB95/

Table 1. Experimental results of algorithms regarding the offline performance

Alg. & Inst.	eil76							
	$f = 20$				$f = 100$			
$m \Rightarrow$	0.1	0.25	0.5	0.75	0.1	0.25	0.5	0.75
S-ACO	403.1	441.1	541.7	752.4	378.5	426.5	505.6	711.5
P-ACO	396.9	437.2	538.5	750.9	381.5	432.3	511.8	723.8
EIIACO	392.5	432.0	532.8	739.3	379.3	428.5	506.4	710.6

Alg. & Inst.	kroA100							
	$f = 20$				$f = 100$			
$m \Rightarrow$	0.1	0.25	0.5	0.75	0.1	0.25	0.5	0.75
S-ACO	18806.3	21106.1	26000.6	35812.8	17726.2	19351.3	23672.0	34232.5
P-ACO	18043.5	20767.0	25777.5	35544.5	17891.0	19656.8	24089.5	34633.1
EIIACO	18041.3	20710.7	25447.4	34963.1	17789.5	19520.1	23800.9	34106.4

Alg. & Inst.	kroA200							
	$f = 20$				$f = 100$			
$m \Rightarrow$	0.1	0.25	0.5	0.75	0.1	0.25	0.5	0.75
S-ACO	25979.4	29058.4	36383.0	50344.3	23519.0	26037.0	33583.6	45338.0
P-ACO	24621.4	28525.8	35924.7	49620.5	23529.9	26284.5	33761.4	45314.5
EIIACO	24686.9	28132.4	35164.3	48190.4	23423.4	26136.9	33302.5	44388.0

Table 2. Statistical tests of comparing algorithms regarding the offline performance

Alg. & Inst.	eil76				kroA100				kroA200			
$f = 20, m \Rightarrow$	0.1	0.25	0.5	0.75	0.1	0.25	0.5	0.75	0.1	0.25	0.5	0.75
S-ACO $\Leftrightarrow$ P-ACO	$s-$	$s-$	$s-$	$s-$	$s-$	$s-$	$s-$	$s-$	$s-$	$s-$	$s-$	$s-$
EIIACO $\Leftrightarrow$ P-ACO	$s+$	$s+$	$s+$	$s+$	$+$	$s+$	$s+$	$s+$	$s-$	$s+$	$s+$	$s+$
EIIACO $\Leftrightarrow$ S-ACO	$s+$	$s+$	$s+$	$s+$	$s+$	$s+$	$s+$	$s+$	$s+$	$s+$	$s+$	$s+$
$f = 100, m \Rightarrow$	0.1	0.25	0.5	0.75	0.1	0.25	0.5	0.75	0.1	0.25	0.5	0.75
S-ACO $\Leftrightarrow$ P-ACO	$s+$	$s+$	$s+$	$s+$	$s+$	$s+$	$s+$	$s+$	$+$	$s+$	$s+$	$-$
EIIACO $\Leftrightarrow$ P-ACO	$s+$	$s+$	$s+$	$s+$	$s+$	$s+$	$s+$	$s+$	$s+$	$s+$	$s+$	$s+$
EIIACO $\Leftrightarrow$ S-ACO	$s-$	$s-$	$-$	$+$	$s-$	$s-$	$s-$	$s+$	$s+$	$s-$	$s+$	$s+$

better understand the dynamic behaviour of the algorithms, the offline performance of the first 500 iterations is plotted in Fig. 1 for fast and slowly changing environments with $m = 0.1$ and $m = 0.75$, respectively. From the experimental results several observations can be made and are analyzed as follows.

First, S-ACO significantly outperforms P-ACO in almost all slowly changing environments whereas it is beaten in fast changing environments; see the results of S-ACO $\Leftrightarrow$ P-ACO in Table 2. This result validates that the S-ACO algorithm can adapt to dynamic changes, but it needs sufficient time to recover and locate the new optimum. The S-ACO algorithm uses pheromone evaporation in order to eliminate pheromone trails that are not useful for the new environment and helps the population of ants to forget the previous solution where they have converged to. On the other hand, P-ACO uses more aggressive method to eliminate previous pheromone trails, which guides the population of ants to keep up with the changing environments, even if they change fast. From Fig. 1, it can be observed that P-ACO converges faster than S-ACO, which helps in fast changing environments but not in slowly changing environments. As we have discussed previously, P-ACO has a high risk to maintain identical ants in the population-list and may get trapped in a local optimum solution.

Second, EIIACO outperforms P-ACO in almost all dynamic test environments; see the results of EIIACO ⇔ P-ACO in Table 2. However, P-ACO is competitive with EIIACO when the environment is slightly changing, e.g., in the kroA200 with $f = 20$ and $m = 0.1$. This is because the solutions stored in the population-list of P-ACO may be fit only when the previous environment has many similarities with the new one. In cases where the environmental changes are medium to significant, the environmental information-based immigrants transfer more knowledge to the pheromone trails of the next iteration and increase the diversity. In slowly changing environments, EIIACO outperforms P-ACO in all dynamic test cases, either with small or large magnitude of changes. This is because EIIACO has enough time to gain knowledge from the previous environment.

Third, EIIACO outperforms S-ACO in all fast changing environments whereas it is competitive in some slowly changing environments; see the results of EIIACO ⇔ S-ACO in Table 2. On the smallest problem instance, i.e., eil76, S-ACO is significantly better than EIIACO because the diversity provided from the immigrants scheme may not be helpful. Furthermore, it is easier for the population in S-ACO to forget previous solutions and become more adaptive. This validates our expectation for S-ACO, where the time needed to adapt depends on the size of the problem and the magnitude of change. As the problem size and magnitude of change increases, EIIACO is significantly better than S-ACO because the population in S-ACO needs more time to adapt in more complex problem instances; see Fig. 1. Moreover, it can be observed that when the magnitude of change is small, S-ACO converges slowly to a better solution, whereas when the magnitude of change is large, EIIACO converges quickly to a better solution.

Finally, in order to investigate the effect of the environmental information-based immigrants scheme in the population diversity of ACO, we calculate the mean population diversity of all iterations as follows:

$$Div = \frac{1}{G} \sum_{i=1}^{G} \left(\frac{1}{N} \sum_{j=1}^{N} \left(\frac{1}{\mu(\mu-1)} \sum_{p=1}^{\mu} \sum_{q \neq p}^{\mu} M_{pq} \right) \right), \qquad (5)$$

where G is the number of iterations, N is the number of runs, μ is the size of the population, $M_{pq} = 1 - \frac{CE(p,q)}{l}$ is the metric that defines the difference between ant p and ant q, where $CE(p,q)$ is the common edges between the ants and l is the number of cities. A value of M_{pq} closer to 0 means that the two ants are similar. The total diversity results for all dynamic test cases are presented in Fig. 2. It can be observed that S-ACO maintains the highest diversity. Especially, in fast changing environments, S-ACO has an extremely high level of diversity, and this

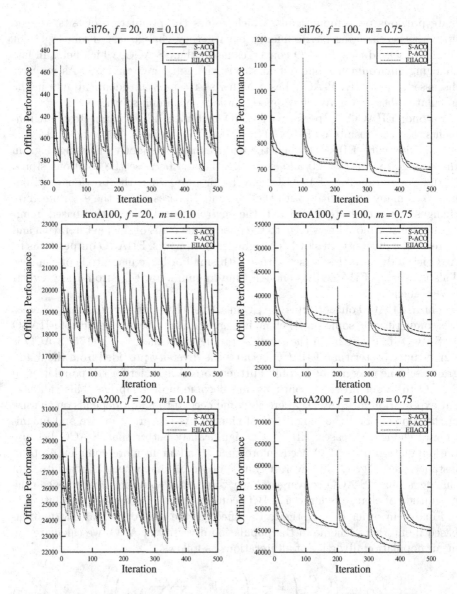

Fig. 1. Dynamic behaviour of investigated algorithms on different DTSPs

shows that it does not have sufficient time to converge. The P-ACO algorithm has the lowest diversity level, which shows the negative effect when identical ants are stored in the population-list. EIIACO maintains higher diversity than P-ACO and much lower diversity than S-ACO. This shows that the diversity generated from the proposed scheme is guided. Moreover, it shows that ACO algorithms that maintain higher diversity levels than others do not always achieve better performance for the DTSP; see Table 1 and Fig. 2.

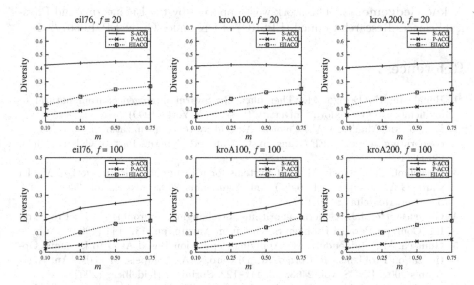

Fig. 2. Overall population diversity of algorithms for different dynamic test problems

6   Conclusions and Future Work

Several immigrants schemes based on individual information, e.g., elitist-based immigrants, have been integrated with ACO algorithms to address different DT-SPs in the literature [8,9]. In this paper, an immigrants scheme based on environmental information, i.e., the frequency of cities appearing next to each other, is proposed for ACO to address the DTSP with traffic factors. A number of generated immigrants replace the worst ants in the current population in order to maintain diversity and transfer knowledge to the pheromone trails for the ants that will construct solutions on the next iteration.

From the experimental results of comparing the proposed EIIACO algorithm with S-ACO and P-ACO algorithms on different cases of DTSPs, the following concluding remarks can be drawn. First, ACO algorithms can be benefited by transferring knowledge from previous environments to the pheromone trails using immigrants schemes for DTSPs. Second, S-ACO has good performance in slowly and slightly changing environments and it is comparable with EIIACO, especially on small problem instances. Third, EIIACO outperforms other ACO algorithms in fast changing environments, while it is comparable with P-ACO in some slightly changing environments. Fourth, EIIACO is significantly better than P-ACO in almost all slowly changing environments. Finally, guided diversity is usually better than random diversity.

For future work, it will be interesting to compare or hybridize EIIACO with other immigrants schemes, which are based on individual information [8,9], and investigate the interaction between the two types of schemes. As another interesting future work, EIIACO can be applied in more challenging optimization problems, e.g., vehicle routing problems [11].

Acknowledgements. This work was supported by the Engineering and Physical Sciences Research Council (EPSRC) of UK under Grant EP/E060722/2.

References

1. Bonabeau, E., Dorigo, M., Theraulaz, G.: Swarm intelligence: from natural to artificial systems. Oxford University Press, New York (1999)
2. Dorigo, M., Maniezzo, V., Colorni, A.: Ant system: optimization by a colony of cooperating agents. IEEE Trans. Syst., Man and Cybern., Part B: Cybern. 26(1), 29–41 (1996)
3. Eyckelhof, C.J., Snoek, M.: Ant Systems for a Dynamic TSP. In: Dorigo, M., Di Caro, G.A., Sampels, M. (eds.) Ant Algorithms 2002. LNCS, vol. 2463, p. 88. Springer, Heidelberg (2002)
4. Grefenestette, J.J.: Genetic algorithms for changing environments. In: Proc. 2nd Int. Conf. on Parallel Problem Solving from Nature, pp. 137–144 (1992)
5. Guntsch, M., Middendorf, M.: Applying Population Based ACO to Dynamic Optimization Problems. In: Dorigo, M., Di Caro, G.A., Sampels, M. (eds.) Ant Algorithms 2002. LNCS, vol. 2463, pp. 111–122. Springer, Heidelberg (2002)
6. Guntsch, M., Middendorf, M.: Pheromone Modification Strategies for Ant Algorithms Applied to Dynamic TSP. In: Boers, E.J.W., Gottlieb, J., Lanzi, P.L., Smith, R.E., Cagnoni, S., Hart, E., Raidl, G.R., Tijink, H. (eds.) EvoWorkshops 2001. LNCS, vol. 2037, pp. 213–222. Springer, Heidelberg (2001)
7. Jin, Y., Branke, J.: Evolutionary optimization in uncertain environments - a survey. IEEE Trans. Evol. Comput. 9(3), 303–317 (2005)
8. Mavrovouniotis, M., Yang, S.: Ant Colony Optimization with Immigrants Schemes in Dynamic Environments. In: Schaefer, R., Cotta, C., Kołodziej, J., Rudolph, G. (eds.) PPSN XI. LNCS, vol. 6239, pp. 371–380. Springer, Heidelberg (2010)
9. Mavrovouniotis, M., Yang, S.: Memory-Based Immigrants for Ant Colony Optimization in Changing Environments. In: Di Chio, C., Cagnoni, S., Cotta, C., Ebner, M., Ekárt, A., Esparcia-Alcázar, A.I., Merelo, J.J., Neri, F., Preuss, M., Richter, H., Togelius, J., Yannakakis, G.N. (eds.) EvoApplications 2011, Part I. LNCS, vol. 6624, pp. 324–333. Springer, Heidelberg (2011)
10. Neumann, F., Witt, C.: Runtime analysis of a simple ant colony optimization algorithm. Algorithmica 54(2), 243–255 (2009)
11. Rizzoli, A.E., Montemanni, R., Lucibello, E., Gambardella, L.M.: Ant colony optimization for real-world vehicle routing problems – from theory to applications. Swarm Intelli. 1(2), 135–151 (2007)
12. Stützle, T., Hoos, H.: The MAX-MIN ant system and local search for the traveling salesman problem. In: Proc. 1997 IEEE Int. Conf. on Evol. Comput., pp. 309–314 (1997)
13. Yang, S.: Genetic algorithms with memory and elitism based immigrants in dynamic environments. Evol. Comput. 16(3), 385–416 (2008)
14. Yu, X., Tang, K., Yao, X.: An immigrants scheme based on environmental information for genetic algorithms in changing environments. In: Proc. 2008 IEEE Cong. of Evol. Comput., pp. 1141–1147 (2008)
15. Yu, X., Tang, K., Chen, T., Yao, X.: Empirical analysis of evolutionary algorithms with immigrants schemes for dynamic optimization. Memetic Comput. 1(1), 3–24 (2009)

A Surrogate-Based Intelligent Variation Operator for Multiobjective Optimization

Alan Díaz-Manríquez, Gregorio Toscano-Pulido, and Ricardo Landa-Becerra

Information Technology Laboratory, CINVESTAV-Tamaulipas
Parque Científico y Tecnológico TECNOTAM
Km. 5.5 carretera Cd. Victoria-Soto La Marina
Cd. Victoria, Tamaulipas 87130, México
{adiazm,gtoscano,rlanda}@tamps.cinvestav.mx

Abstract. Evolutionary algorithms are meta-heuristics that have shown flexibility, adaptability and good performance when solving Multiobjective Optimization Problems (MOPs). However, in order to achieve acceptable results, Multiobjective Evolutionary Algorithms (MOEAs) usually require several evaluations of the optimization function. Moreover, when each of these evaluations represents a high computational cost, these expensive problems remain intractable even by these meta-heuristics. To reduce the computational cost in expensive optimization problems, some researchers have replaced the real optimization function with a computationally inexpensive surrogate model. In this paper, we propose a new intelligent variation operator which is based on surrogate models. The operator is incorporated into a stand-alone search mechanism in order to perform its validation. Results indicate that the proposed algorithm can be used to optimize MOPs. However, it presents premature convergence when optimizing multifrontal MOPs. Therefore, in order to solve this drawback, the proposed operator was successfully hybridized with a MOEA. Results show that this latter approach outperformed both, the former proposed algorithm and the evolutionary algorithm but without the operator.

Keywords: Evolutionary Algorithms, Intelligent Genetic Variation Operator, Multiobjective Optimization.

1 Introduction

Evolutionary Algorithms (EAs) have been successfully applied to an important variety of difficult multiobjective optimization problems (MOPs). EAs are population-based techniques that make a multidimensional search, finding more than one solution within an execution. Their main advantage lies on their ability to locate solutions close to the global optimum even in highly accident search spaces. However, for many real-world optimization problems, the number of calls of the objective function to locate a suboptimal solution may be high. In many science and engineering problems, researchers have been using computer simulation in order to replace expensive physical experiments with the aim of improving the quality and performance of engineered products and devices, but using

J.-K. Hao et al. (Eds.): EA 2011, LNCS 7401, pp. 13–24, 2012.

a fraction of the needed effort. For example, Computational Fluid Dynamics solvers, Computational Electro Magnetics and Computational Structural Mechanics have been shown to be very accurate. However, such simulations are often computationally expensive, such that, some simulations can take several days or even weeks in order to be completed. Hence, in many real-world optimization problems, the computational cost is dominated by the quantity of objective function evaluations needed to obtain a good solution. This suggests that it is crucial to use any information gained during the optimization process in order to increase the efficiency of optimization algorithms and consequently, to reduce the number of objective function evaluations.

A natural approach used by the EA community to reduce the number of evaluations has been the use of a surrogate model instead of the exact problem [9,5,2]. Building a model of the fitness function to assist in the selection of candidate solutions for evaluation has been a natural approach used by the EA community since it reduces the number of evaluations [9,5,2].

An alternative approach to reduce the number of evaluations needed to obtain suboptimal solutions by an EA is to use information from the local search space in order to generate several promising solutions for the next generation, these techniques are known as **Intelligent Genetic Operators** [3,1,11,10,6]

Hence, in this work, we propose a new intelligent genetic variation operator that uses a surrogate model to guide the search with the aim of accelerating its convergence i.e., achieve similar results but using a lower number of function evaluations. The remainder of this work is organized as follows: In Section 2, we give a brief background needed for the understanding of the paper. Section 3 presents the proposed intelligent genetic variation operator denominated Surrogate-based Intelligent Variation Operator (SIVO) and we validate it using a simple approach called Surrogate Assisted Optimizer (SAO). Section 4 presents the incorporation of SIVO into the NSGA-II. Experimental results obtained on synthetic test problems are presented in Section 5. Finally, Section 6 summarizes our main conclusions as well as some possible directions for future work.

2   Background

Definition 1 (Multiobjective Optimization Problem - MOP): A multiobjective optimization problem can be defined as: Find the vector $x^* = [x_1^*, x_2^*, ..., x_n^*]^T \in \mathcal{F}$ which will satisfy the m inequality constrains, $g_i(x) \leq 0$ $i = 1, 2, ..., m$, the p equality constrains $h_j(x) = 0$ $j = 1, 2, ..., p$ and optimize the vector function: $f(x) = [f_1(x), f_2(x), ..., f_k(x)]^T$, where $x = [x_1, x_2, ..., x_n]^T$ is the vector of decision variables, and the set of inequality and equality constrains will define the feasible region $\mathcal{F}$ and any point in $\mathcal{F}$ will constitute a feasible solution.

Definition 2 (Pareto Optimality): A point $x^* \in \mathcal{F}$ is **Pareto-optimal** if every $x \in \mathcal{F}$ and $I = \{1, 2, ..., k\}$ either, $\forall_{i \in I}(f_i(x) = f_i(x^*))$ or, there is at least one $i \in I$ such that: $f_i(x) > f_i(x^*)$.

Definition 3 (Pareto Dominance): A vector $x = [x_1, x_2, ..., x_n]^T$ is said to dominate $y = [y_1, y_2, ..., y_n]^T$ (denoted by $x \preceq y$ if and only if x is partially less than y, i.e., $\forall_{i \in \{1,...,n\}} : x_i \leq y_i \land \exists_{i \in \{1,...,n\}} : x_i < y_i$.

Definition 4 (Pareto Optimal Set): For a given MOP $f(x)$, the Pareto optimal set $(\mathcal{P}^*)$ is defined as: $\mathcal{P}^* = \{x \in \mathcal{F} \mid \nexists y \in \mathcal{F} \mid f(y) \preceq f(x)\}$

Definition 5 (Pareto Front): For a given MOP $f(x)$, and Pareto optimal set $\mathcal{P}^*$, the Pareto front $(\mathcal{PF}^*)$ is defined as:
$\mathcal{PF}^* = \{f(x) = [f_1(x), f_2(x), ..., f_k(x)]^T \mid x \in \mathcal{P}^*\}$

Definition 6 (Surrogate Model): A surrogate model is a mathematical model that mimics the behavior of a computationally expensive simulation code over the complete parameter space as accurately as possible, using the least amount of data points.

3 Surrogate-Based Intelligent Variation Operator (SIVO)

The Surrogate-based Intelligent Variation Operator **SIVO** can be an alternative or a complement to the common genetic variation operators. A key feature which we want for **SIVO** is to improve the solutions previously found by the generation of new non-dominated solutions. For this purpose, we first select a non-dominated solution in order to improve it. Then, we propose three new (desired) solutions in the objective space which outperform the selected one. After that, we try to locate their positions in the variable space. Therefore, we construct the local surrogate model and launch a new optimization algorithm, which tries to find the surrogated solutions which are closest to the three previously proposed points. At the end, the final surrogate solutions are evaluated in the real optimization problem with the aim that the new solutions found be better than the original one.

3.1 Selection in the Objective Space of the New Solutions to Be Created

The fundamental idea of **SIVO** is that given a solution x and its evaluation $f(x)$, we want to find new solutions (S) whose evaluation $(f(S))$ are situated at a distance Δf. For a bi-objective problem, the evaluation of the solutions $f(S)$ that we wish to find are shown graphically in Figure 1, such solutions can be calculated as follows:

$$f''(S_1) = [f_1(x) - \Delta f_1 \quad f_2(x)] \tag{1}$$

$$f''(S_2) = [f_1(x) \quad f_2(x) - \Delta f_2] \tag{2}$$

$$f''(S_3) = [f_1(x) + \Delta f_1 \quad f_2(x) - 2\Delta f_2] \tag{3}$$

It is obvious that solutions $f''(S_1)$ and $f''(S_2)$ will help us to improve the algorithm's convergence with respect to the original point. Moreover, $f''(S_3)$ will

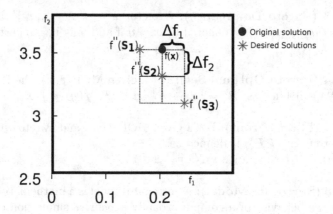

Fig. 1. Solutions that we want to create with the operator, $\Delta f = [\Delta f_1 \quad \Delta f_2]$

help to improve the diversity with respect to $f(\boldsymbol{x})$. Finally, we can distinguish that $f''(S_1)$, $f''(S_2)$ and $f''(S_3)$ are not dominated among them.

3.2   Creation and Optimization of the Local Surrogate Model

Once we have our targets (the three non-dominated points in the objective space that we would like to find), then we proceed to create the surrogate model with the solutions previously stored.

Once we know the position of the desired solution we would like to find $S = [S_1 \ S_2 \ S_3]$ where:

$$f(S_1) \approx f''(S_1) \wedge f(S_2) \approx f''(S_2) \wedge f(S_3) \approx f''(S_3) \qquad (4)$$

Then we proceed with the creation of a local surrogate model with any surrounding solutions previously stored. It should be obvious that we need to store every evaluated solution in a repository in order to use such information for building the surrogate models. Therefore, in order to select the surrounding solutions, we use a simple clustering algorithm, such that, we can find the solutions which belong to the same cluster of $\boldsymbol{x}$ (the clustering algorithm is performed in the variable space). After we have the surrounding solutions, then we create a surrogate model $f'(x)$ with such solutions. After that, we minimize the distance of the surrogate model to each solution in S. The formulation of the optimization problem for each solution i in S is as follows:

Find x' such that,

$$\text{Minimize} \sum_{j=1}^{M}(f'_j(\boldsymbol{x}'_i) - f''_j(S_i))^2 \qquad (5)$$

where M refers to the number of objectives of the MOP. In this work, we use the Nelder-Mead algorithm to optimize the above optimization problem. The solutions x_i' found will be the new solutions created by the operator $NS = \{S_1, S_2, S_3\}$. We recommend to optimize the previous optimization problem with a local search algorithm and use the solution x as initial solution, because in this way if the optimizer can not find an exact solution, at least we can get a solution close to x. Although the error (Equation 6) is a important factor, there is not a strong effect in **SIVO**, because if we create new solutions with **SIVO** and they have a big error with respect to the original point, then, in the next generation these solutions created will serve as new solutions for the creation of the surrogate model, and therefore the error would be lower. The pseudocode for **SIVO** is shown in Algorithm 1.

$$error = \sum_{1=1}^{|S|} \mathrm{abs}(f(S_i) - f''(S_i)) \qquad (6)$$

Algorithm 1. SIVO$(x, \Delta f, P)$

Input: Solution x, Δf, Stored solutions P
Output: New solutions NS
 $NS = \emptyset$
 $f''(S_1) \leftarrow [f_1(x) - \Delta f_1 \quad f_2(x)]$
 $f''(S_2) \leftarrow [f_1(x) \quad f_2(x) - \Delta f_2]$
 $f''(S_3) \leftarrow [f_1(x) + \Delta f_1 \quad f_2(x) - 2\Delta f_2]$
 $C \leftarrow Clustering(P)$
 $S_C \leftarrow \{\forall_{i \in P} C_i = C_x\}$ {C_x is the Cluster which belong x}
 $f' = $Metamodel$(Sc)$ {The surrogate model is created with S_c}
 for $i \leftarrow 1$ to 3 **do**
  $x_i' \leftarrow$Solve problem in Equation 5 with $f''(S_i)$
  $NS \leftarrow NS \cup x_i'$
 end for

3.3   Surrogate Assisted Optimizer (SAO)

In order to evaluate the effectiveness of the proposed operator, we decided to build a simple stand-alone algorithm based on SIVO. The resulting algorithm was called the Surrogate Assisted Optimizer or SAO for short. SAO works as follows: first an initial population of solutions P_0 of size $|P|$ is created using a latin hypercube[1] [4]. Then, it starts the main procedure of SAO which refers to the selection of the solutions that SIVO will try to improve. Therefore, we select β solutions in P according to their dominance rank and their crowding distance. Then the SIVO algorithm is applied to those solutions. The resulting solutions from this procedure will be stored in P. Since P is fixed-size archive, then dominance rank and crowding distance will be applied again in order to maintain the file size. The SAO pseudocode is show in Algorithm 2.

---

[1] A latin hypercube is a statistical method for stratified sampling that can be applied to multiple variables.

Algorithm 2. SAO Algorithm

Input: $|P|$ (size of population), $tmax$ (number of iterations), β (percentage of solutions to apply
   SIVO, $\beta = [0, |P_0|])$, Δf
Output: P_{tmax} (Final population)
1: $P_0 \leftarrow$ GenerateRandomPopulation()
2: **for** $t \leftarrow 0$ to $tmax$ **do**
3:     $S \leftarrow$ Choose β solutions of P_t (with dominance rank and crowding distance)
4:     **for** $j \leftarrow$ to $|S|$ **do**
5:         $NewSolutions \leftarrow$ **SIVO**$(S_j, \Delta f, P_t)$
6:         $NS \leftarrow NS \cup NewSolutions$
7:     **end for**
8:     $P_t \leftarrow P_t \cup NS$
9:     $F \leftarrow$ fast-non-dominated-sorting(P_t)
10:    $P_{t+1} = \emptyset$
11:    **repeat**
12:        Assign crowding distance on F_i
13:        $P_{t+1} \leftarrow P_{t+1} \cup F_i$
14:    **until** $|P_{t+1}| + |F_i| \leq |P_t|$
15:    Crowding distance on F_i
16:    $P_{t+1} \leftarrow P_{t+1} \cup$(Choose the first $(|P| - |P_{t+1}|)$ individuals of F_i)
17: **end for**

3.4   SAO Results

In order to allow a quantitative assessment of the performance of the proposed algorithms, $ZDT\{1-4,6\}$ test problems [12] and two metrics were adopted.

1. **Inverted Generational Distance (IGD):** The concept of generational distance was introduced by Van Veldhuizen & Lamont [7,8] as a way of estimating how far are the elements in the Pareto front produced by our algorithm from those in the true Pareto front of the problem. This measure is defined as:

$$GD = \frac{\sqrt{\sum_{i=1}^{n} d_i^2}}{n} \tag{7}$$

   where n is the number of non-dominated vectors found by the algorithm being analyzed and d_i is the Euclidean distance (measured in objective space) between each of these and the nearest member of the true Pareto front. It should be clear that a value of $GD = 0$ indicates that all the elements generated are in the true Pareto front of the problem. Therefore, any other value will indicate how "far" we are from the global Pareto front of our problem. In our case, we implemented the "inverted" generational distance measure (IGD) in which we use as a reference the true Pareto front, and we compare each of its elements with respect to the front produced by an algorithm. In this way, we are calculating how far are the elements of the true Pareto front, from those in the Pareto front produced by our algorithm. In this metric, we will prefer smaller values of IGD, since this represents a closer result to the real front.

2. **Hypervolume (HV)** The HV calculates the area covered by all the solutions in Q using a reference point W, which is a vector identified as having the worst objective function values.

$$HV = volume(\bigcup_{i=1}^{|Q|} v_i) \tag{8}$$

where for each solution $i \in Q$, a hypercube v_i, is constructed with reference to W. The HV is the union of all hypercubes. The HV depends strongly of the choice of W. A set of non-dominated solutions can be different values for two different choices of W. In this metric, we will prefer larger values of HV, since this represents a closer result to the real front.

The parameters used for **SAO** are a initial population of $|P| = 100$, a $\Delta f = [0.1\ 0.5]$, and $\beta = 5$. With a $\beta = 5$, the algorithm by generation will do 15 evaluations, because the **SIVO** creates 3 new solutions by point, therefore, $\beta*3 = 15$. We measure the hypervolume at every 100 evaluations during 2000 function evaluations in our approach. This procedures was executed during 31 times and the mean of the results was plotted.

Figure 2 shows the performance of **SAO** in the five test functions adopted. In all the problems, we can see that the hypervolume had a rapid growth such that, **SAO** required only (approx.) 400 evaluations to obtain good results. However, since the HV does not tell us whether the algorithm is close to the real front, we decided to use the IGD performance measure. IGD confirms, in most of the problems (ZDT1, ZDT2, ZDT3, and ZDT5), the results obtained by HV, due to SAO obtained an IGD very close to 0 with only 400 evaluations. But, when optimizing the ZDT4 test function, the IGD reached its best value within 700 evaluations, and from this point the IGD improvement was practically null.

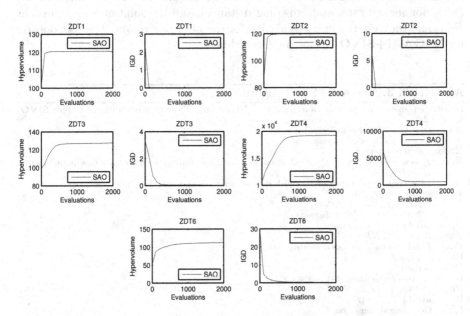

Fig. 2. Hypervolume and Inverted Generational Distance for SAO

Moreover, this value is relatively far from 0 (where 0 refers that the solutions found are on the true Pareto front). This result shows that the algorithm tends to stuck in local optima in this type of problems (multifrontal problems). This allows us to speculate that an hybridization of **SIVO** with an Evolutionary Algorithm (EA), can improve the overall results.

4   The NSGA-II + SIVO

SAO could work properly when trying to optimize most of the adopted problems. However, it was also obvious that SAO did not behave as it was expected when it tried to optimize multifrontal problems, since it presented premature convergence. Therefore, it was clear that an additional effort in order to develop a more "robust" algorithm was necessary. In order to solve such a problem, we decided hybridize the **SIVO** with an Evolutionary Algorithm. In this section, we describe the methodology used for the integration of **SIVO** with the **NSGA-II**.

The **NSGA-II+SIVO** works as follows: First, it creates a random initial population of solutions P_0 of size $|P|$ (in this work, the initial population was created with latin hypercubes). Then, in the iteration t a child population Q_t is created with the traditional genetic operators (mutation and crossover), a mixed population R_t is created with $P_t \cup Q_t$, the solutions in R_t are used as stored solutions for **SIVO**. It is easy to see that the main procedure of the integration of **NSGA-II** with **SIVO** is the choice of the solutions to apply the operator, we choose β solutions in R_t to apply **SIVO**, the solutions are chosen according to their dominance rank and their crowding distance (as in NSGA-II), the new solutions created NS are mixed with R_t and its size is limited to $|P|$ (with dominance rank and crowding distance), the $|P|$ solutions are stored in P_{t+1}, and they are the new population for the next generation. The pseudocode for **NSGA-II+SIVO** is show in Algorithm 3.

Algorithm 3. NSGA-II+SIVO Algorithm

Input: $|P|$ (size of population), $tmax$ (number of iterations), β (number of solutions to apply **SIVO**, β), Δf

Output: P_{tmax} (Final population)

1: $P_0 \leftarrow$ GenerateRandomPopulation()
2: **for** $t \leftarrow 0$ to $tmax$ **do**
3:    $Q_t \leftarrow$ make-new-pop(P_t){use selection, crossover and mutation to create a new population}
4:    $R_t \leftarrow P_t \cup Q_t$
5:    $S \leftarrow$ Choose β solutions of R_t (with dominance rank and crowding distance)
6:    **for** $j \leftarrow$ to $|S|$ **do**
7:       $NewSolutions \leftarrow$ **SIVO**($S_j, \Delta f, R_t$)
8:       $NS \leftarrow NS \cup NewSolutions$
9:    **end for**
10:   $R_t \leftarrow R_t \cup NS$
11:   $F \leftarrow$ fast-non-dominated-sorting(R_t)
12:   $P_{t+1} = \emptyset$
13:   **repeat**
14:      Assign crowding distance on F_i
15:      $P_{t+1} \leftarrow P_{t+1} \cup F_i$
16:   **until** $|P_{t+1}| + |F_i| \leq |P_t|$
17:   Crowding distance on F_i
18:   $P_{t+1} \leftarrow P_{t+1} \cup$(Choose the first ($|P| - |P_{t+1}|$) individuals of F_i)
19: **end for**

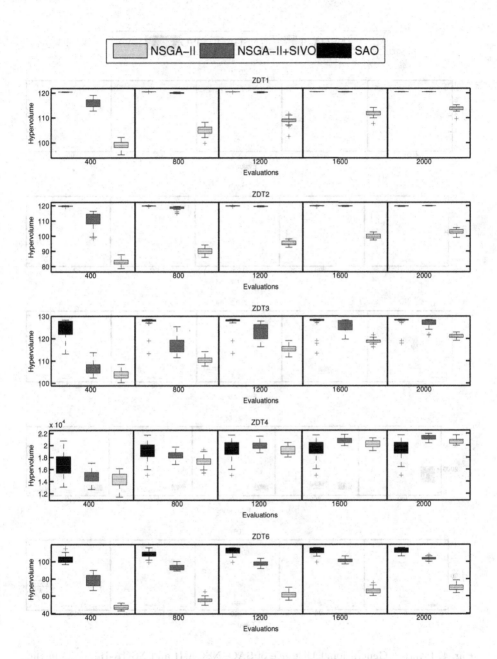

Fig. 3. Hypervolume of SAO, NSGA-II and NSGA-II+SIVO in the five test problems

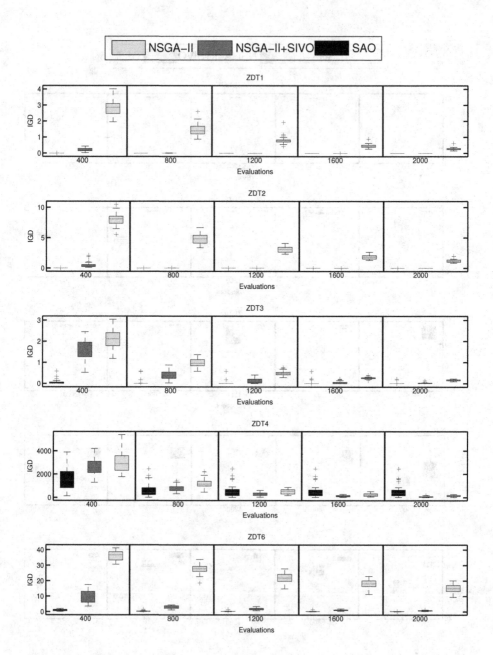

Fig. 4. Inverted Generational Distance of SAO, NSGA-II and NSGA-II+SIVO in the five test problems

5 Comparison of Results

We selected the following parameters for NSGA-II and NSGA-II+SIVO: $|P| = 100$, and for NSGA-II+SIVO were used a $|P| = 84$, a $\Delta f = [0.1\ 0.5]$, and $\beta = 5$. Both algorithms used a mutation rate of $1/Number\_of\_variables$, crossover rate of 0.8, an index crossover of 15, and a mutation index of 20. We also decided to compare with respect to our previously proposed approach (SAO), therefore, we selected the same parameters as indicated before. In order to know how the behavior of the proposed approach is, we applied the hypervolume and IGD performance measures each 400 evaluations during 2000 evaluations. This procedure was executed 31 times and the results were plotted as boxplot graphics.

Figure 3 shows the results of hypervolume in NSGA-II, NSGA-II+SIVO and SAO. As pointed out above, SAO had troubles to escape from local optima in multifrontal problems (ZDT4). While NSGA-II+SIVO clearly shows how the algorithm achieves a significant improvement with respect to the increase of the number of evaluations. Moreover, comparing both, the original NSGA-II and the NSGA-II+SIVO, it is clear that the SIVO operator helped to improve the convergence rate of NSGA-II+SIVO.

The IGD results are shown in the Figure 4, in most of the problems the SAO algorithm obtained good results. However, it can be seen in this metric that only the hybrid approach (NSGA-II+SIVO) could deal with the ZDT4 test problem.

6 Conclusions

The purpose of this paper was to propose a new operator based on surrogate models. The operator can be used as a simple stand-alone optimizer, or it can be added to an evolutionary algorithm, such that, we propose both schemes, a stand-alone optimizer: The Surrogate Assisted Optimizer (SAO), and an evolutionary algorithm: The NSGA-II+SIVO. The first algorithm proposed has the property to have a rapid convergence, but it also can be trapped in a false front when optimizing multifrontal problems. For this reason, we developed the second algorithm, this algorithm maintain the speed of convergence of SAO, but also incorporates the exploratory capability of NSGA-II. After comparing this algorithm with respect to the NSGA-II, we found that the proposed algorithm could outperform both, the original NSGA-II and SIVO stand alone algorithm. Further investigation and experimentation on the hybridization of the SIVO with other evolutionary algorithms, and an extension of SIVO for three objective problems are needed. Also, we are planning to work in the automated calculation of δ.

Acknowledgments. The first author acknowledges support from CONACyT through a scholarship to pursue graduate studies at the Information Technology Laboratory, CINVESTAV-Tamaulipas. The second author gratefully acknowledges support from CONACyT through project 105060.

References

1. Anderson, K.S., Hsu, Y.: Genetic Crossover Strategy Using an Approximation Concept. In: Angeline, P.J., Michalewicz, Z., Schoenauer, M., Yao, X., Zalzala, A. (eds.) Proceedings of the 1999 Congress on Evolutionary Computation, CEC 1999, Washington, D.C., USA, vol. 1, p. 533 (1999)
2. Emmerich, M., Giannakoglou, K., Naujoks, B.: Single and Multiobjective Evolutionary Optimization Assisted by Gaussian Random Field Metamodels. IEEE Transactions on Evolutionary Computation 10(4), 421–439 (2006)
3. Jun, S., Shigenobu, K.: Extrapolation-Directed Crossover for Real-coded GA: Overcoming Deceptive Phenomena by Extrapolative Search. In: Proceedings of the 2001 Congress on Evolutionary Computation (CEC 2001), Seoul, Korea, vol. 1, pp. 655–662. IEEE Press (2001)
4. McKay, M.D., Beckman, R.J., Conover, W.J.: A Comparison of Three Methods for Selecting Values of Input Variables in the Analysis of Output from a Computer Code. Technometrics 21(2), 239–245 (1979)
5. Santana-Quintero, L.V., Serrano-Hernandez, V.A., Coello, C.A., Hernández-Díaz, A.G., Molina, J.: Use of Radial Basis Functions and Rough Sets for Evolutionary Multi-Objective Optimization. In: Proceedings of the 2007 IEEE Symposium on Computational Intelligence in Multicriteria Decision Making (MCDM 2007), Honolulu, Hawaii, USA, pp. 107–114. IEEE Press (April 2007)
6. Talukder, A.K.M.K.A., Kirley, M., Buyya, R.: The Pareto-Following Variation Operator as an Alternative Approximation Model. In: IEEE Congress on Evolutionary Computation, CEC 2009, pp. 8–15 (May 2009)
7. Veldhuizen, D.A.V., Lamont, G.B.: Multiobjective Evolutionary Algorithm Research: A History and Analysis. Technical Report TR-98-03, Department of Electrical and Computer Engineering, Graduate School of Engineering, Air Force Institute of Technology, Wright-Patterson AFB, Ohio (1998)
8. Veldhuizen, D.A.V., Lamont, G.B.: On Measuring Multiobjective Evolutionary Algorithm Performance. In: 2000 Congress on Evolutionary Computation, Piscataway, New Jersey, vol. 1, pp. 204–211. IEEE Service Center (July 2000)
9. Li, Q., Liao, X., Yang, X., Zhang, W., Li, W.: Multiobjective optimization for crash safety design of vehicles using stepwise regression model. Structural and Multidisciplinary Optimization 35, 561–569 (2008)
10. Zhang, Q., Sun, J., Tsang, E.: An Evolutionary Algorithm with Guided Mutation for the Maximum Clique Problem. IEEE Transactions on Evolutionary Computation 9(2), 192–200 (2005)
11. Zhou, Q., Li, Y.: Directed Variation in Evolution Strategies. IEEE Transactions on Evolutionary Computation 7(4), 356–366 (2003)
12. Zitzler, E., Deb, K., Thiele, L.: Comparison of Multiobjective Evolutionary Algorithms: Empirical Results. Evolutionary Computation 8(2), 173–195 (2000)

The Relationship between the Covered Fraction, Completeness and Hypervolume Indicators

Viviane Grunert da Fonseca[1,3] and Carlos M. Fonseca[2,3]

[1] INUAF – Instituto Superior D. Afonso III, Loulé, Portugal
`viviane.grunert@sapo.pt`
[2] CISUC, Department of Informatics Engineering
University of Coimbra, Coimbra, Portugal
`cmfonsec@dei.uc.pt`
[3] CEG-IST – Center for Management Studies
Instituto Superior Técnico, Lisbon, Portugal

Abstract. This paper investigates the relationship between the covered fraction, completeness, and (weighted) hypervolume indicators for assessing the quality of the Pareto-front approximations produced by multiobjective optimizers. It is shown that these unary quality indicators are all, by definition, weighted Hausdorff measures of the intersection of the region attained by such an optimizer outcome in objective space with some reference set. Moreover, when the optimizer is stochastic, the indicators considered lead to real-valued random variables following particular probability distributions. Expressions for the expected value of these distributions are derived, and shown to be directly related to the first-order attainment function.

Keywords: stochastic multiobjective optimizer, performance assessment, covered fraction indicator, completeness indicator, (weighted) hypervolume indicator, attainment function, expected value, Hausdorff measure.

1 Introduction

The performance assessment of stochastic multiobjective optimizers (MOs) has become an emerging area of research, enabling the comparison of existing optimizers and supporting the development of new ones. In general, stochastic MO performance can be associated with the distributional behavior of the random outcomes produced in one optimization run, seen either as random non-dominated point (RNP) sets in objective space or, alternatively, as the corresponding random unbounded *attained* sets, also in objective space [7].

Typically, realizations of such random closed sets in $\mathbb{R}^d$ can be observed arbitrarily often through *multiple* optimization runs. This allows stochastic MO performance assessment and comparison to be carried out with frequency-based statistical inference methodology using (simple) random samples of independent and identically distributed MO outcome sets.

J.-K. Hao et al. (Eds.): EA 2011, LNCS 7401, pp. 25–36, 2012.
© Springer-Verlag Berlin Heidelberg 2012

However, employing the frequency argument for MO outcome sets is not an easy task. Often, the overall outcome-set distribution is very complex – too complex to be considered as a whole in most practical situations. Therefore, in order to formulate suitable estimators and/or hypothesis tests for MO performance assessment and comparison, it needs to be agreed upon what *partial* aspect of this set distribution one is interested in.

Currently, two main stochastic MO performance assessment approaches are in use, but without really exploiting the relationship between them: the *attainment function approach* [8,5,7] and the *quality indicator approach* [17,15].

The first approach describes the optimizer outcome distribution directly via a hierarchy of nested, increasingly informative attainment functions, where all functions beyond a certain order lead to a full performance description. It is known that the first-order attainment function relates to the *location* and *spread* of an outcome-set distribution, while second and higher-order versions address the corresponding *inter-point dependence structures* [7]. However, to the authors' knowledge, third- and higher-order attainment functions have not yet been considered in practice, due to computational difficulties [5,6].

The quality indicator approach addresses the complexity of the outcome-set distribution in a different way, through the definition of so-called (unary) quality indicators which somehow transform each realized outcome set into a scalar. The corresponding, much simpler, univariate distributions of indicator values are usually studied via an estimate of their expected value (average indicator value), although other statistics could be considered as well. Even though this approach clearly implies a loss of information about the overall MO performance, there is hardly any discussion in the literature about what aspect of the outcome-set distribution is, or is not, addressed by each indicator.

On the whole, it is generally difficult to combine or compare results of different MO performance studies, unless exactly the same assessment methodology is used. Little is known about which indicators can complement each other and which, when used together, supply redundant information. Hence, there is a need to classify quality indicators with respect to the information they provide. Acknowledging the fact that unary quality indicators are transformations of the original optimizer outcome set, it seems natural to explore this link and attempt to relate the resulting indicator-value distributions with the attainment function hierarchy. As a first step in this direction, the present paper discusses the covered fraction, the completeness and the (weighted) hypervolume indicators with respect to their definition and to the mean (or expected value) of the corresponding distributions.

In Section 2, the attainment function and the quality indicator approaches are briefly outlined. Subsequently, the covered fraction indicator, the completeness indicator and the (weighted) hypervolume indicator are considered in Sections 3 to 5, respectively. The paper ends with a discussion of the results in Section 6 and some concluding remarks in Section 7.

2   Approaches to MO Performance Assessment

In the context of performance assessment, the outcome set of a multiobjective optimizer is considered to be the image in objective space of the non-dominated solutions produced in one optimization run (obeying some stopping criterion). When the optimizer is stochastic, such a set of objective vectors is random, and its probability distribution reflects the performance of the optimizer on a given optimization problem instance.

Mathematically, the outcome set of a d-objective stochastic optimizer can be interpreted as a random closed set [11]. More specifically, it is a random non-dominated point set (RNP set) [7]

$$\mathcal{X} = \{X_1, \ldots, X_M \in \mathrm{I\!R}^d : \; P(X_i \leq X_j) = 0, \; i \neq j\}, \qquad (1)$$

where both the cardinality M and the elements X_i, $i = 1, \ldots, M$, are random, and $P(0 \leq M < \infty) = 1$. In other words, $\mathcal{X}$ has a finite, but random, number of elements which do not weakly dominate one another in the Pareto sense [17]. Therefore, stochastic MO performance is related to

1. the (identical) multivariate distribution of the random vectors $X_1, \ldots, X_M$,
2. the way in which these random vectors depend on each other (in pairs, triples, quadruples, etc.) and, finally,
3. the univariate distribution of the discrete random variable M.

For simplicity, it has been common practice to condition on M with realizations up to a certain value m^* and/or to study only *partial* aspects of the set distribution of $\mathcal{X}$. The two main approaches in this context are outlined below.

2.1   Attainment Function Approach

Let m^* be the maximum number of non-dominated objective vectors in $\mathrm{I\!R}^d$ that may be generated by a d-objective optimizer. Then, increasing amounts of distributional information about the corresponding outcome set $\mathcal{X}$ with growing $k = 1, \ldots, m^*$ are comprised in the attainment functions $\alpha_{\mathcal{X}}^{(k)} : \mathrm{I\!R}^{d \times k} \longrightarrow [0, 1]$, where

$$\alpha_{\mathcal{X}}^{(k)}(z_1, \ldots, z_k) = P(\mathcal{X} \trianglelefteq z_1 \wedge \ldots \wedge \mathcal{X} \trianglelefteq z_k), \qquad (2)$$

and the event

$$[\mathcal{X} \trianglelefteq z] \quad \Longleftrightarrow \quad [X_1 \leq z \vee X_2 \leq z \vee \ldots \vee X_M \leq z] \qquad (3)$$

denotes the attainment of a goal $z \in \mathrm{I\!R}^d$ by $\mathcal{X}$, assuming minimization without loss of generality. Thus, (2) gives the probability that the outcome set $\mathcal{X}$ weakly dominates the set of goals $\{z_1, \ldots, z_k\}$. The random unbounded but closed set $\mathcal{Y} = \{z \in \mathrm{I\!R}^d : \mathcal{X} \trianglelefteq z\}$, which contains all goals $z \in \mathrm{I\!R}^d$ dominated by at least one element of $\mathcal{X}$, is known as the *attained set* [7].

A *complete* distributional performance description (given that $M \leq m^*$) is provided by the attainment function of order $k = m^*$, while the first-order attainment function $\alpha_{\mathcal{X}}^{(1)}(\cdot) = \alpha_{\mathcal{X}}(\cdot)$ is sufficient to characterize where in objective space goals tend to be attained (*location*) and with what degree of variability this happens across multiple runs (*spread*).

The above (theoretical) attainment functions can be estimated from a random sample of independent and identically distributed RNP sets $\mathcal{X}_1, \mathcal{X}_2, \ldots, \mathcal{X}_n$ via the cumulative frequencies of the corresponding (combinations of) events of the type $[\mathcal{X} \trianglelefteq z]$. For $k = 1$, for example, the non-parametric estimator is the *empirical first-order attainment function*

$$\alpha_n(z) = \frac{1}{n} \sum_{i=1}^n \mathbf{I}\{z \in \mathcal{Y}_i\} = \frac{1}{n} \sum_{i=1}^n \mathbf{I}\{\mathcal{X}_i \trianglelefteq z\} , \tag{4}$$

where $\mathbf{I}_A(\cdot) = \mathbf{I}\{\cdot \in A\}$ denotes the indicator function of the set A defined over $\mathbb{R}^d$. Estimators for higher-order attainment functions can be constructed in a similar way [7].

2.2   Quality Indicator Approach

Here, the outcome RNP set $\mathcal{X}$ is transformed into a single real-valued random variable $I(\mathcal{X})$, usually with respect to some non-empty, deterministic, closed reference set $Z_{ref} \subset \mathbb{R}^d$. A variety of such set-transformations are currently in use, each of which defining a particular unary quality indicator that reflects stochastic MO performance, in a restricted sense, via the associated *univariate* probability distribution of indicator values.

Like the distribution of $\mathcal{X}$, these indicator (value) distributions are unknown for a given MO application. Arising from the transformation applied to $\mathcal{X}$, they depend on the distribution of $\mathcal{X}$ itself, which can be described through the attainment function hierarchy. Hence, determining the form of this dependence should reveal what aspect(s) of the optimizer outcome-set distribution each indicator actually addresses.

Indicator distributions can, in principle, be estimated as a whole using the information of n independent optimization runs, in a non-parametric way. For simplicity, however, it is common to begin by estimating their expected values $\mathrm{E}[I(\mathcal{X})]$ through the sample average $\frac{1}{n} \sum_{i=1}^n I(\mathcal{X}_i)$. Similarly, this work focuses on the expected value of the distributions produced by the quality indicators considered.

3   Covered Fraction Indicator

The original version of the (unary) covered fraction indicator, or coverage indicator, considers the proportion of the Pareto-optimal front $\mathcal{X}^*$ (assumed to be a finite point set in $\mathbb{R}^d$) that is attained by the outcome set $\mathcal{X}$ [13,15]. In this work, a more general definition will be used which, instead of referring to $\mathcal{X}^*$, may refer to any deterministic, non-empty, compact (i.e., closed and bounded) reference set Z_{ref} in objective space $\mathbb{R}^d$.

3.1   Definition

Given a general subset $A \subset \mathbb{R}^d$, denote its (normalized) *Hausdorff measure* of dimension j by $\mathcal{H}^j(A)$, where $j \geq 0$. Provided that A is non-empty, there is a single value of j, known as the *Hausdorff dimension* of A, for which $\mathcal{H}^j(A)$ is finite and positive [1,12, p. 105]. Then, $\mathcal{H}^j(A)$ may be understood as the "size" of A. For example:

- $\mathcal{H}^0(\cdot)$, also known as the *counting measure*, measures the cardinality of a point set in $\mathbb{R}^d$ (where any such set has Hausdorff dimension zero),
- $\mathcal{H}^1(\cdot)$ measures the length of a Hausdorff one-dimensional smooth curve in $\mathbb{R}^d$ (for example a straight line, a circle or an ellipse),
- $\mathcal{H}^2(\cdot)$ measures the area of a Hausdorff two-dimensional smooth surface[1] in $\mathbb{R}^d$ (for example a plane or the surface of a sphere), and finally
- $\mathcal{H}^d(\cdot)$ measures the hypervolume of a Hausdorff d-dimensional set in $\mathbb{R}^d$, and, as such, corresponds to the usual Lebesgue measure on $\mathbb{R}^d$.

Thus, taking a Hausdorff j-dimensional reference set $Z_{ref} \subset \mathbb{R}^d$, $0 \leq j \leq d$, the covered fraction indicator of $\mathcal{X}$ can be generally defined as[2]

$$I_{CF}(\mathcal{X}, Z_{ref}) = \frac{\mathcal{H}^j(\{z \in Z_{ref} : \ \mathcal{X} \trianglelefteq z\})}{\mathcal{H}^j(Z_{ref})} \tag{5}$$

$$= \frac{1}{\displaystyle\int_{\mathbb{R}^d} \mathbf{I}\{z \in Z_{ref}\} \, \mathcal{H}^j(dz)} \cdot \int_{Z_{ref}} \mathbf{I}\{\mathcal{X} \trianglelefteq z\} \, \mathcal{H}^j(dz) \, . \tag{6}$$

For Hausdorff d- and zero-dimensional reference sets in objective space $\mathbb{R}^d$, the above definition (with integrals with respect to the measure $\mathcal{H}^j$) can be written in more familiar ways, as follows:

1. For a reference set $Z_{ref} \subset \mathbb{R}^d$ of Hausdorff dimension d, the covered fraction indicator of $\mathcal{X}$ can be defined as a quotient of Lebesgue integrals:

$$I_{CF}(\mathcal{X}, Z_{ref}) = \frac{1}{\displaystyle\int_{\mathbb{R}^d} \mathbf{I}\{z \in Z_{ref}\} \, dz} \cdot \int_{Z_{ref}} \mathbf{I}\{\mathcal{X} \trianglelefteq z\} \, dz \, . \tag{7}$$

---

[1] A j-dimensional smooth surface is the image of a continuously differentiable mapping (i.e. of class C^1) from $\mathbb{R}^j$ onto $\mathbb{R}^d$, where $j < d$ [9, p. 355]. Note that a countable union of Hausdorff j-dimensional sets preserves the Hausdorff dimension [4, p. 112], i.e. any such union of smooth sets is also of integer Hausdorff dimension. A non-mathematical discussion of the concept of "smoothness" can be found in [3, p. 335].

[2] The Hausdorff dimension of any realization of the outcome set $\mathcal{X}$ is equal to zero, and the Hausdorff dimension of a non-empty *attained set* realization is equal to d. Non-smooth reference sets, such as fractal sets, may have a non-integer Hausdorff dimension.

2. For a discrete reference set $Z_{ref}^k = \{z_1, \ldots, z_k\}$ of k not necessarily non-dominated points in $\mathbb{R}^d$, the definition simplifies to

$$I_{CF}(\mathcal{X}, Z_{ref}^k) = \frac{1}{k} \cdot \sum_{j=1}^{k} \mathbf{I}\{\mathcal{X} \trianglelefteq z_j\} . \tag{8}$$

In any case, the covered fraction indicator of $\mathcal{X}$ takes realizations in $[0, 1]$, where a larger observed indicator value is considered to represent a "better" optimization result (in the particular sense of the indicator).

3.2 Expected Value

Due to the "linearity property of expectation", and since the binary random variable $\mathbf{I}\{\mathcal{X} \trianglelefteq z\}$ follows a Bernoulli distribution with expected value $P(\mathcal{X} \trianglelefteq z) = \alpha_{\mathcal{X}}(z)$, for all $z \in \mathbb{R}^d$, it can be seen that the expected value of the covered fraction indicator distribution is related to the first-order attainment function. For a Hausdorff j-dimensional reference set Z_{ref}, $0 \leq j \leq d$, it holds that

$$E[I_{CF}(\mathcal{X}, Z_{ref})] = \frac{E[\mathcal{H}^j(\{z \in Z_{ref} : \mathcal{X} \trianglelefteq z\})]}{\mathcal{H}^j(Z_{ref})} \tag{9}$$

$$= \frac{1}{\int_{\mathbb{R}^d} \mathbf{I}\{z \in Z_{ref}\} \, \mathcal{H}^j(dz)} \cdot \int_{Z_{ref}} \alpha_{\mathcal{X}}(z) \, \mathcal{H}^j(dz) , \tag{10}$$

where (10) can be obtained by applying Robbins's theorem [11, p. 59] to the random compact set $\{z \in Z_{ref} : \mathcal{X} \trianglelefteq z\}$, since $\mathcal{H}^j$ is *locally finite* on a Hausdorff j-dimensional Z_{ref}.

For Hausdorff d- and zero-dimensional reference sets, the above formula can again be simplified:

1. For a reference set $Z_{ref} \subset \mathbb{R}^d$ of Hausdorff dimension d,

$$E[I_{CF}(\mathcal{X}, Z_{ref})] = \frac{1}{\int_{\mathbb{R}^d} \mathbf{I}\{z \in Z_{ref}\} \, dz} \cdot \int_{Z_{ref}} \alpha_{\mathcal{X}}(z) \, dz . \tag{11}$$

2. For a discrete reference set $Z_{ref}^k = \{z_1, \ldots, z_k\}$ of Hausdorff dimension zero,

$$E[I_{CF}(\mathcal{X}, Z_{ref}^k)] = \frac{1}{k} \cdot \sum_{j=1}^{k} \alpha_{\mathcal{X}}(z_j) . \tag{12}$$

4   Completeness Indicator

In its original form, the completeness indicator is defined for a given (observed) solution set in decision space (after one optimization run), as the probability of selecting a point uniformly at random from the feasible set which is weakly dominated by that solution set [10,15].

4.1   Definition

For a random outcome set $\mathcal{X}$ in objective space, the original completeness indicator can be defined as the conditional probability of $\mathcal{X}$ attaining a non-uniformly distributed random reference point V from the image of the feasible set, given $\mathcal{X}$. However, the indicator may also be defined more generally by considering some deterministic, non-empty, closed reference set $Z_{ref} \subset \mathbb{R}^d$ (which may or may not be the image of the feasible set), together with a random vector V taking realizations in Z_{ref} and following an explicitly defined distribution. Hence,

$$I_{CO}(\mathcal{X}, V, Z_{ref}) = P\left(\mathcal{X} \trianglelefteq V \mid \mathcal{X}\right) . \tag{13}$$

Like the covered fraction indicator of $\mathcal{X}$, the completeness indicator of $\mathcal{X}$ is a random variable which takes realizations in $[0, 1]$, where larger values correspond to "better" optimization results.

In fact, it can be seen that the definitions of the two indicators are very similar, when interpreting $I_{CO}(\mathcal{X}, V, Z_{ref})$ as the conditional expectation of $\mathbf{I}\{\mathcal{X} \trianglelefteq V\}$ given $\mathcal{X}$: with a Hausdorff j-dimensional reference set $Z_{ref} \subset \mathbb{R}^d$, $0 \leq j \leq d$, and a random vector V supported on Z_{ref}, stochastically *independent* from the outcome set $\mathcal{X}$, and for which the probability density function $f_V(\cdot)$ exists with respect to Hausdorff measure $\mathcal{H}^j$, it holds that

$$I_{CO}(\mathcal{X}, V, Z_{ref}) = P\left(\mathcal{X} \trianglelefteq V \mid \mathcal{X}\right) \tag{14}$$

$$= \mathrm{E}\left[\mathbf{I}\{\mathcal{X} \trianglelefteq V\} \mid \mathcal{X}\right] \tag{15}$$

$$= \int_{Z_{ref}} \mathbf{I}\{\mathcal{X} \trianglelefteq z\} \cdot f_{V|\mathcal{X}}(z) \, \mathcal{H}^j(dz) \tag{16}$$

$$= \int_{Z_{ref}} \mathbf{I}\{\mathcal{X} \trianglelefteq z\} \cdot f_V(z) \, \mathcal{H}^j(dz) . \tag{17}$$

Clearly, the particular choice of a *uniform* distribution for $V = V_u$ over some compact reference set Z_{ref}, again stochastically independent from $\mathcal{X}$, leads to the identity of the covered fraction indicator and the completeness indicator. In this case,

$$f_{V_u}(z) = f_u(z) = 1/\mathcal{H}^j(Z_{ref}) \tag{18}$$

for all $z \in Z_{ref}$, and $f_u(z) = 0$ otherwise.

Finally, note that for both Hausdorff d-dimensional reference sets and zero-dimensional reference sets, the formulation in (17) can be simplified, respectively in the sense of (7) and (8), by substituting the Hausdorff measure $\mathcal{H}^j$ either by the usual Lebesgue measure on $\mathbb{R}^d$ or by the counting measure for point sets.

4.2   Expected Value

From (17) and the argumentation given in subsection 3.2, it immediately follows that

$$\mathrm{E}[I_{CO}(\mathcal{X}, V, Z_{ref})] = \int_{Z_{ref}} \alpha_{\mathcal{X}}(z) \cdot f_V(z) \, \mathcal{H}^j(dz) , \tag{19}$$

while more familiar formulations can be achieved for Hausdorff d- and zero-dimensional reference sets:

1. For a reference set $Z_{ref} \subset \mathbb{R}^d$ of Hausdorff dimension d,

$$\mathrm{E}[I_{CO}(\mathcal{X}, V, Z_{ref})] = \int_{Z_{ref}} \alpha_{\mathcal{X}}(z) \cdot f_V(z) \, dz \,. \tag{20}$$

2. For a discrete reference set $Z_{ref}^k = \{z_1, \ldots, z_k\}$ of Hausdorff dimension zero,

$$\mathrm{E}[I_{CO}(\mathcal{X}, V, Z_{ref}^k)] = \sum_{j=1}^{k} \alpha_{\mathcal{X}}(z_j) \cdot P(V = z_j) \,. \tag{21}$$

5   Hypervolume Indicator

The hypervolume indicator (also known as dominated space), introduced in [16], considers the size of the portion of a deterministic, non-empty, closed reference set in objective space $\mathbb{R}^d$ that is attained by the outcome set $\mathcal{X}$. Later, this idea was refined by including a weight function that assigns varying levels of importance to different regions of the reference set [14].

In contrast to the covered fraction and completeness indicators, the reference set for the (weighted) hypervolume indicator is usually specified through a single reference point $z_{ref}^* \in \mathbb{R}^d$, and can be written as

$$Z_{ref}^* = \{z \in \mathbb{R}^d : \ z \leq z_{ref}^*\} \,, \tag{22}$$

though in some cases the Pareto-optimal front $\mathcal{X}^*$ or some other set of non-dominated points below z_{ref}^* is used to bound Z_{ref}^* from below, in such a way that it is still Hausdorff d-dimensional.

5.1   Definition

For a reference set Z_{ref}^* as given in (22), the (weighted) hypervolume indicator is defined as

$$I_H(\mathcal{X}, w, Z_{ref}^*) = \int_{Z_{ref}^*} \mathbf{I}\{\mathcal{X} \trianglelefteq z\} \cdot w(z) \, dz \,, \tag{23}$$

where $w(\cdot)$ is a non-negative valued weight function, integrable over Z_{ref}^*, i.e.

$$\int_{Z_{ref}^*} w(z) \, dz < \infty \,. \tag{24}$$

In this general form, the (weighted) hypervolume indicator of $\mathcal{X}$ takes realizations in the interval

$$\left[0, \int_{Z_{ref}^*} w(z) \, dz\right] \,, \tag{25}$$

where larger observed indicator values are associated with "better" optimization results.

Some choices for the weight function $w(\cdot)$ deserve special attention:

– The indicator weight function, $w(z) = \mathbf{I}_{Z^*_{ref}}(z), z \in \mathbb{R}^d$, defines the classical, non-weighted hypervolume indicator with possible values ranging from 0 to the size of Z^*_{ref}, i.e., $\mathcal{H}^d(Z^*_{ref})$, if the reference set is compact.
– When $w(\cdot)$ is the *probability density* function [2] of a random variable V,

$$\int_{Z^*_{ref}} w(z)\, dz = 1\,, \tag{26}$$

and all possible indicator values are contained in the interval $[0, 1]$. Further, if V is independent from $\mathcal{X}$, the (weighted) hypervolume indicator corresponds to the completeness indicator with respect to the reference set Z^*_{ref}.
– For the probability density function $w(\cdot) = f_u(\cdot)$ of a random variable V_u, *uniformly* distributed over some compact reference set Z^*_{ref} and independent from $\mathcal{X}$, all three indicators considered in this paper are identical, i.e.

$$I_{CF}(\mathcal{X}, Z^*_{ref}) = I_{CO}(\mathcal{X}, V_u, Z^*_{ref}) = I_H(\mathcal{X}, f_u, Z^*_{ref})\,. \tag{27}$$

5.2 Expected Value

Again, from the argumentation given in subsection 3.2, it follows that

$$E[I_H(\mathcal{X}, w, Z^*_{ref})] = \int_{Z^*_{ref}} \alpha_{\mathcal{X}}(z) \cdot w(z)\, dz\,. \tag{28}$$

6 Discussion

The covered fraction, completeness and weighted hypervolume indicators are all, by definition, weighted Hausdorff measures of the intersection of the attained set $\mathcal{Y} = \{z \in \mathbb{R}^d : \mathcal{X} \trianglelefteq z\}$ with some reference set Z_{ref} in objective space $\mathbb{R}^d$. As such, they can be seen as special cases of a *generalized hypervolume indicator* of $\mathcal{X}$, defined as

$$I_{GH}(\mathcal{X}, w, Z_{ref}) = \int_{Z_{ref}} \mathbf{I}\{\mathcal{X} \trianglelefteq z\} \cdot w(z)\, \mathcal{H}^j(dz)\,, \tag{29}$$

where

– Z_{ref} is a Hausdorff j-dimensional, deterministic, closed reference set in $\mathbb{R}^d$, $0 \leq j \leq d$, representing the region in objective space which is of interest for performance assessment, and
– $w(\cdot)$ is a non-negative valued weight function, integrable over Z_{ref}, which assigns different levels of importance to the points in Z_{ref}.

Note that, technically, the weight function $w(\cdot)$ alone would be sufficient to parametrize this indicator, as its support would define the corresponding reference set Z_{ref}. In other words, the generalized hypervolume indicator measures

Table 1. Indicators as special cases of the *generalized hypervolume indicator*

Indicator	Reference set	Weight function
weighted hypervolume	Z_{ref}^*, closed	integrable over Z_{ref}^*
non-weighted hypervolume	Z_{ref}^*, compact	indicator function $\mathbf{I}_{Z_{ref}^*}(\cdot)$
completeness	Z_{ref}, closed	density function f_V with support on Z_{ref}, V independent of $\mathcal{X}$
covered fraction	Z_{ref}, compact	uniform density function with support on Z_{ref}

the *mass* of the region attained by an optimizer outcome set, as determined by some *mass density* function $w(\cdot)$, which may or may not be also a probability density function. The generalized hypervolume indicator can be reduced to each of the indicators considered in this paper, as summarized in Table 1.

As a function of the outcome set $\mathcal{X}$, the indicator $I_{GH}(\mathcal{X}, w, Z_{ref})$ leads to a random variable with realizations in the interval

$$\left[0, \int_{Z_{ref}} w(z)\, \mathcal{H}^j(dz)\right], \tag{30}$$

where larger indicator values indicate a "better" optimization result. At this point it should be noted, however, that for the purpose of "unit-independent" comparisons of optimizer performance, it may be preferable to always use a normalizing density weight function, so that indicator values can be limited to the interval $[0, 1]$.

For a general weight function $w(\cdot)$, the expected value of the generalized hypervolume indicator distribution can be expressed as

$$\mathrm{E}[I_{GH}(\mathcal{X}, w, Z_{ref})] = \int_{Z_{ref}} \alpha_{\mathcal{X}}(z) \cdot w(z)\, \mathcal{H}^j(dz). \tag{31}$$

For the *probability density* function of a random vector V, supported on Z_{ref} and distributed *independently* from the outcome set $\mathcal{X}$, this additionally leads to the interesting relationship:[3]

$$\mathrm{E}[I_{GH}(\mathcal{X}, f_V, Z_{ref})] = \mathrm{E}[\alpha_{\mathcal{X}}(V)] = P(\mathcal{X} \trianglelefteq V). \tag{32}$$

Hence, the expected value of the generalized hypervolume indicator, and consequently of all three indicators considered in this paper, can be obtained from the first-order attainment function.

This result is important because it sheds light on the relationship between the quality indicator approach and the attainment function approach. It can be seen that the three indicators considered, unified in the generalized hypervolume indicator, convey information related to the *location* of the optimizer outcome-set

---

[3] For independent $\mathcal{X}$ and V: $\mathrm{E}\left[P(\mathcal{X} \trianglelefteq V \mid V)\right] = \mathrm{E}\left[P(\mathcal{X} \trianglelefteq V \mid \mathcal{X})\right] = P(\mathcal{X} \trianglelefteq V)$.

distribution via their mean indicator values. Moments of the generalized hypervolume indicator distribution other than the mean may still contain information beyond that captured by the first-order attainment function.

The generalized hypervolume indicator also allows new indicators to be constructed using less usual reference sets and/or weight functions. For example, assume that the outcome set $\mathcal{X}$ of a hypothetical two-objective optimizer, when applied to a given problem instance, is simply the set of minima [6] of a set of two stochastically independent random vectors in $\mathbb{R}^2$. Assume also that both vectors are distributed according to a bivariate exponential distribution with parameter $\lambda > 0$, with density function

$$f(t_1, t_2) = \lambda^2 \cdot e^{-\lambda \cdot t_1 - \lambda \cdot t_2} \cdot \mathbf{I}_{[0,\infty)^2}((t_1, t_2)') \tag{33}$$

and cumulative distribution function

$$F(t_1, t_2) = \left(1 - e^{-\lambda \cdot t_1} - e^{-\lambda \cdot t_2} + e^{-\lambda \cdot t_1 - \lambda \cdot t_2}\right) \cdot \mathbf{I}_{[0,\infty)^2}((t_1, t_2)') . \tag{34}$$

Then, the first-order attainment function of $\mathcal{X}$ at a goal $z = (t_1, t_2)' \in \mathbb{R}^2$ is

$$\alpha_{\mathcal{X}}(z) = 1 - \left(1 - F(t_1, t_2)\right)^2 . \tag{35}$$

Given $(r_1, r_2)' \in \mathbb{R}^2_+$, the reference set $Z_{seg} = \{(r_1 \cdot t, r_2 - r_2 \cdot t)' : t \in [0, 1]\}$ is a line segment in $\mathbb{R}^2$, and has Hausdorff dimension 1. In this case, the generalized hypervolume indicator with weight function $\mathbf{I}_{Z_{seg}}(\cdot)$ measures the *length* of the intersection of this segment with the attained set. From (31), the expected indicator value may then be calculated via the line integral

$$\mathrm{E}[I_{GH}(\mathcal{X}, \mathbf{I}_{Z_{seg}}, Z_{seg})] = \sqrt{r_1^2 + r_2^2} \cdot \int_0^1 \alpha_{\mathcal{X}}((r_1 \cdot t, r_2 - r_2 \cdot t)') \, dt . \tag{36}$$

Clearly, the expected value can also be directly evaluated using the standard integral based on the joint density of the two exponential random vectors, but the latter leads to much more complicated expressions.

7 Concluding Remarks

In this paper the relationship between three popular quality indicators for multiobjective optimizer performance assessment has been studied with respect to their definitions. By considering Hausdorff measures, a *single* notation was introduced for all indicators, leading to their unification. Furthermore, it was shown that, when the optimizer is stochastic, the corresponding expected indicator values depend on the first-order attainment function of the optimizer outcomes.

Investigating how the variance and other aspects of these univariate indicator-value distributions relate to the attainment function hierarchy will be the subject of future work, as well as considering other unary quality indicators. In particular, it would be interesting to see how indicators designed to assess the dispersion or uniformity of points in observed approximation sets relate in distribution to (higher-order) attainment functions.

References

1. Alberti, G.: Geometric measure theory. In: Françoise, J.P., et al. (eds.) Encyclopedia of Mathematical Physics, vol. 2, pp. 520–527. Elsevier, Oxford (2006)
2. Bader, J.M.: Hypervolume-Based Search for Multiobjective Optimization: Theory and Methods. Ph.D. thesis, Swiss Federal Institute of Technology, Zurich (2009)
3. Barenblatt, G.I.: Scaling, Self-similarity, and Intermediate Asymptotics. Cambridge University Press, Cambridge (1996)
4. DiBenedetto, E.: Real Analysis. Birkhäuser, Boston (2002)
5. Fonseca, C.M., Grunert da Fonseca, V., Paquete, L.: Exploring the Performance of Stochastic Multiobjective Optimisers with the Second-Order Attainment Function. In: Coello Coello, C.A., Hernández Aguirre, A., Zitzler, E. (eds.) EMO 2005. LNCS, vol. 3410, pp. 250–264. Springer, Heidelberg (2005)
6. Fonseca, C.M., Guerreiro, A.P., López-Ibáñez, M., Paquete, L.: On the Computation of the Empirical Attainment Function. In: Takahashi, R.H.C., Deb, K., Wanner, E.F., Greco, S. (eds.) EMO 2011. LNCS, vol. 6576, pp. 106–120. Springer, Heidelberg (2011)
7. Grunert da Fonseca, V., Fonseca, C.M.: The attainment-function approach to stochastic multiobjective optimizer assessment and comparison. In: Bartz-Beielstein, T., et al. (eds.) Experimental Methods for the Analysis of Optimization Algorithms, ch. 5, pp. 103–130. Springer, Berlin (2010)
8. Grunert da Fonseca, V., Fonseca, C.M., Hall, A.O.: Inferential Performance Assessment of Stochastic Optimisers and the Attainment Function. In: Zitzler, E., Deb, K., Thiele, L., Coello Coello, C.A., Corne, D. (eds.) EMO 2001. LNCS, vol. 1993, pp. 213–225. Springer, Heidelberg (2001)
9. Ito, K. (ed.): Encyclopedic Dictionary of Mathematics 2. The Mathematical Society of Japan. The MIT Press (1987)
10. Lotov, A.V., Bushenkov, V.A., Kamenev, G.K.: Interactive Decision Maps: Approximation and Visualization of Pareto Frontier. Kluwer Academic Publishers, Dordrecht (2004)
11. Molchanov, I.: Theory of Random Sets. Springer, London (2005)
12. Ott, E.: Chaos in Dynamical Systems. Cambridge University Press, Cambridge (2002)
13. Ulungu, E.L., Teghem, J., Fortemps, P.H., Tuyttens, D.: MOSA method: A tool for solving multiobjective combinatorial optimization problems. Journal of Multi-Criteria Decision Analysis 8(4), 221–236 (1999)
14. Zitzler, E., Brockhoff, D., Thiele, L.: The Hypervolume Indicator Revisited: On the Design of Pareto-compliant Indicators Via Weighted Integration. In: Obayashi, S., Deb, K., Poloni, C., Hiroyasu, T., Murata, T. (eds.) EMO 2007. LNCS, vol. 4403, pp. 862–876. Springer, Heidelberg (2007)
15. Zitzler, E., Knowles, J., Thiele, L.: Quality Assessment of Pareto Set Approximations. In: Branke, J., Deb, K., Miettinen, K., Słowiński, R. (eds.) Multiobjective Optimization. LNCS, vol. 5252, pp. 373–404. Springer, Heidelberg (2008)
16. Zitzler, E., Thiele, L.: Multiobjective Optimization Using Evolutionary Algorithms - A Comparative Case Study. In: Eiben, A.E., Bäck, T., Schoenauer, M., Schwefel, H.-P. (eds.) PPSN 1998. LNCS, vol. 1498, pp. 292–301. Springer, Heidelberg (1998)
17. Zitzler, E., Thiele, L., Laumanns, M., Fonseca, C.M., Grunert da Fonseca, V.: Performance assessment of multiobjective optimizers: An analysis and review. IEEE Transactions on Evolutionary Computation 7(2), 117–132 (2003)

A Rigorous Runtime Analysis for Quasi-Random Restarts and Decreasing Stepsize

Marc Schoenauer, Fabien Teytaud, and Olivier Teytaud

TAO (Inria), LRI, UMR 8623 (CNRS - Univ. Paris-Sud), bat 490 Univ.
Paris-Sud 91405 Orsay, France

Abstract. Multi-Modal Optimization (MMO) is ubiquitous in engineering, machine learning and artificial intelligence applications. Many algorithms have been proposed for multimodal optimization, and many of them are based on restart strategies. However, only few works address the issue of initialization in restarts. Furthermore, very few comparisons have been done, between different MMO algorithms, and against simple baseline methods. This paper proposes an analysis of restart strategies, and provides a restart strategy for any local search algorithm for which theoretical guarantees are derived. This restart strategy is to decrease some 'step-size', rather than to increase the population size, and it uses quasi-random initialization, that leads to a rigorous proof of improvement with respect to random restarts or restarts with constant initial step-size. Furthermore, when this strategy encapsulates a (1+1)-ES with 1/5th adaptation rule, the resulting algorithm outperforms state of the art MMO algorithms while being computationally faster.

1 Introduction

The context of this work is continuous black-box Multi-Modal Optimization (MMO). Given an objective function F, the goal of MMO is to discover **all** global optima of F (and not just a few of them). Because of the "black-box" hypothesis, this paper focuses on gradient-free methods, more particularly Evolution Strategies (ES) [5], addressing both theoretical and experimental issues.

Note that there are to-date very few rigorous analyses of ES for MMO. But it has been experimentally shown that in the MMO setting, choosing a large population size λ reduces the risk being trapped in local minima [26,9,1]. Unfortunately, it has also been shown that some classical ES are not very efficient for large λ [4], and that self-adaptive ES are quite fast and reliable in that case, with a good speed-up as a function of λ [4].

Several MMO-specific methods have been proposed in the evolutionary framework, and their precise description cannot be included here for space considerations. Please refer to the survey proposed in [22] (or to the original papers of course). A popular family of MMO algorithms is based on niching techniques: sharing [8]; clearing [21], including the modified clearing approach proposed in [22]; crowding [6], including deterministic [17] and probabilistic [18] versions; clustering [29]; species conserving genetic algorithms; and finally, different restart

J.-K. Hao et al. (Eds.): EA 2011, LNCS 7401, pp. 37–48, 2012.
© Springer-Verlag Berlin Heidelberg 2012

strategies, to the recent state-of-the-art restart with increasing population size [1]. Several other works have been devoted to MMO outside the evolutionary community. For instance, EGO [12], IAGO [28] provide very efficient algorithms in terms of precision/number of fitness evaluations, but are computationally far too expensive unless the objective function is itself very costly (requiring hours or days of computation per point). Moreover, they have no theoretical guarantee in spite of their elegant derivation. UNLEO [3] proposed an approximation of optimal optimization algorithm under robustness constraints, with nice theoretical guarantees; however, it is, too, far too expensive, and could hence be tested with a few tenths of function evaluations.

This paper proposes a very simple restart strategy that can encapsulate any (local) optimization algorithm, and for which convergence rates can be theoretically derived. This strategy is based on quasi-random restarts, and a decreasing schedule for some 'step-size' parameter. Comparative experiments with the extensive results published in [22] are used to demonstrate the applicability and efficiency of the proposed strategy, in the case where the embedded local algorithm is a simple (1+1)-ES with 1/5th adaptation rule. However, the proposed restart strategy heavily relies on a specific parameter d (see Alg. 1), somehow analogous to a parameter used in the modified clearing which outperformed by far all other algorithms [22]. Hence, because this parameter d will be tuned for the problems used in the experiments, no fair comparison can be done with other published results for which no such parameter tuning was performed. Our conclusions will therefore be limited to the design of a generic restart algorithm based on quasi-random sequences with theoretical guarantees and tight complexity bounds (see Corollary 1), that outperforms uniform restarts, and reaches state of the art performance provided a good setting of parameter d, which essentially quantifies some prior knowledge about the minimal distance between two optima.

The remaining of the paper is organized in the following way: Section 2 introduces the notations used throughout the paper, and surveys the state-of-the-art in quasi-random (QR) sequences. Section 3 introduces the proposed restart strategy, and rigorously quantifies the improvement provided by QR points in the initialization of MMO and the requirement that the initial step-size after a restart goes to zero as the number of restarts increases. Section 4 provides comparative experimental results with state-of-the-art MMO methods as well as the simple random restart method with constant step-size.

2   Mathematical Background

Notations: For each E, subset of a topological space, $\overline{E}$ will denote the *topological closure* of E, i.e., the intersection of all closed sets containing E, and $\#\mathbb{E}$ denote the number of elements of E. For each sequence $S = s_1, s_2, s_3, \ldots$ of points in a topological space, the *accumulation* of S is defined as $Acc\ S = \cap_{n \geq 1}\{s_{n+1}, s_{n+2}, \ldots\}$. For each sequence $X = x_1, x_2, \ldots$ of points in $[0,1]^D$, the *dispersion* of X is defined as $Disp(X, n) = \sup_{x \in [0,1]^D} \inf_{i \in [[1,n]]} ||x - x_i||$.

Quasi-Random (QR) Points. A quasi-random sequence is a (possibly randomized) sequence with some uniformity properties that make it, intuitively, "more uniform" than a pseudo-random sequence. After astonishing results in numerical integration (convergence in $1/n$ instead of $1/\sqrt{n}$ for numerical integration, within logarithmic factors) and successful experiments in random searchs [19], QR points have been used in several works dealing with evolution strategies, during initialization [13,7] or mutation [2,25,24]. Furthermore, "modern" QR sequences using scrambling [14] have been demonstrated to outperform older QR sequences [25,24]. Hence, following [27,24], this paper will only consider Halton sequences with random scrambling, and this Section will only briefly introduce them – please refer to [19,20] for more details and references.

Let us first define Van Der Corput's sequence: Given a prime number p, the nth element $vdc_{n,p}$ of the Van Der Corput sequence in basis p is defined by:

- write n in basis p: $n = d_k d_{k-1} \ldots d_1$, i.e. $n = \sum_{i=0}^{k} d_i p^i$ with $d_i \in [[0, p-1]]$;
- $vdc_{n,p} = 0.d_1 d_2 \ldots d_k$ in basis p, i.e. $vdc_{n,p} = \sum_{i=1}^{k} d_i p^{-i} \in [0,1]$.

A classical improvement in terms of discrepancy for moderate values of n and large values of d, termed *scrambling*, defines $vdc_{n,p}$ as $vdc_{n,p} = 0.\pi(d_1)\pi(d_2)\ldots\pi(d_k)$ where π is some permutation of $[[0, p-1]]$ such that $\pi(0) = 0$ in order to ensure $\forall n, vdc_{n,p} \neq 0$.

Halton sequences generalize Van Der Corput sequences to dimension D by using one different prime number per dimension. Let $p_i, i \in [[1, D]]$ be D prime numbers. The nth element h_n of a Halton sequence in dimension D is defined by $h_n = (vdc_{n,p_1}, vdc_{n,p_2}, \ldots, vdc_{n,p_D}) \in [0,1]^D$. Scrambled-Halton sequences, like the ones used in this paper, are scrambled using a randomly drawn permutation for each $i \in [[1, D]]$.

The Nth Hammersley point set is $\{\text{HAMM}_{N,1}, \text{HAMM}_{N,2}, \ldots, \text{HAMM}_{N,N}\}$, where $\text{HAMM}_{N,n} = ((n-1)/N, vdc_{n,p_1}, vdc_{n,p_2}, \ldots, vdc_{n,p_{D-1}})$.

3   Theoretical Analysis of a Simple Restart Strategy

Let $\mathcal{D}$ be the optimization domain, embedded in a normed vector space (with norm $||.||$). Let $\mathcal{F}$ be a family of fitness functions, i.e., of mappings from $\mathcal{D}$ to $\mathbb{R}$. For any fitness function $f \in \mathcal{F}$, let $X^*(f)$ be the set of interest[1]. Let ES be an optimization algorithm that takes as input $x \in \mathcal{D}$ and $\sigma > 0$ (initial point and step-size respectively), depends on f, and outputs some $x' = ES(x, \sigma, f) \in \mathcal{D}^2$. Think of σ as a radius at which local optima are searched; however, it is not necessary to assume that ES always finds an optimum x' at distance $< \sigma$ of x, but only that for σ sufficiently small and x sufficiently close to x^*, $x' = ES(x, \sigma, f) = x^*$. Finally, for any given sequence $S = (x_i, \sigma_i)_{i \in \mathbb{N}} \in (\mathcal{D} \times]0, \infty[)^{\mathbb{N}}$, denote $RS(S)$ the restart algorithm that successively starts from (x_1, σ_1), (x_2, σ_2), ...

---

[1] In most cases, $X^*(f)$ will be the set of local optima of f, though the results below have some generality w.r.t. $X^*(f)$.

[2] x' is the output of a whole run of ES: in general, it will be the best point of the run; however, here again the results are more general.

Definitions

(i) *ES* has the **convergence property** if

$$(\forall f \in \mathcal{F}) \, (\forall (x, \sigma) \in \mathcal{D} \times]0, \infty[) \, (ES(x, \sigma, f) \in X^*(f)) \tag{1}$$

(ii) ES has the **locality property** w.r.t. $\mathcal{F}$ if

$$(\forall x^* \in X^*(f)) \, (\exists \epsilon > 0) \, (\exists \sigma_0 > 0) \text{ s.t.}$$
$$(\|x - x^*\| < \epsilon) \text{ and } (0 < \sigma < \sigma_0) \Rightarrow (ES(x, \sigma, f) = x^*). \tag{2}$$

Note that this does not imply that $ES(x, \sigma, f^*)$ is necessarily within distance σ of x, which would be an unrealistic assumption.

(iii) ES has the **strong locality property** w.r.t. $\mathcal{F}$ if

$$(\exists \epsilon > 0) \, (\exists \sigma_0 > 0) \text{ s.t. } (\forall f \in \mathcal{F}) \, (\forall x^* \in X^*(f))$$
$$(\|x - x^*\| < \epsilon) \text{ and } (0 < \sigma < \sigma_0) \Rightarrow ES(x, \sigma, f) = x^*. \tag{3}$$

(iv) For any sequence $S \in (\mathcal{D} \times]0, \infty[)^{\mathbb{N}}$, the restart algorithm $RS(S)$ is said to be **consistent w.r.t.** $\mathcal{F}$ if

$$(\forall f \in \mathcal{F}) \, \{ES(x_i, \sigma_i, f); i \in \mathbb{N}\} = X^*(f).$$

We can now state the following consistency theorem.

Theorem 1 (Consistency of the restart algorithm). *Assume that ES has the convergence property (Eq. 1) and the locality property (Eq. 2). Then:*

1. *If $(X^*(f) \times \{0\} \subset Acc \, S)$ for all $f \in \mathcal{F}$, then $RS(S)$ is consistent w.r.t. $\mathcal{F}$;*
2. *If $\mathcal{D} \times \{0\} \subset Acc \, S$, then $RS(S)$ is consistent w.r.t. all $\mathcal{F} \subset \mathbb{R}^{\mathcal{D}}$.*

Proof: (2) is an immediate consequence of (1) so let us prove (1).
Let $f \in \mathcal{F}$. Eq. 1 immediately implies that $\{ES(x_i, \sigma_i, f); i \geq 1\} \subset X^*(f)$.
Let $x^* \in X^*(f)$; then $(x^*, 0)$ is in $Acc \, S$ by assumption. Using Eq. 2, it follows that $x^* = ES(x_i, \sigma_i, f)$ for some $i \geq 1$; this proves that $X^*(f) \subset \{ES(x_i, \sigma_i, f); i \geq 1\}$; hence the expected result. □

Remark 1. The assumption $X^*(f) \times \{0\} \subset Acc \, S$ is necessary. It holds in particular with random restarts points and random initial step-sizes with non-zero density close to 0; or random restart points and step-sizes decreasing to 0; in both cases, quasi-random restarts can be used instead of random restarts.

Next result now considers the number of restarts required for finding all points in $X^*(f)$. Sequences of starting points are now considered stochastic, hence the expectation operator:

Proposition 1 (QR restarts are faster). *Assume that ES has the convergence property (Eq. 1) and the strong locality property (Eq. 3) for some ϵ, σ_0. Assume that $\sigma_i = \sigma_0$ for all $i \geq 1$. Define*
$$\#RS(\mathcal{F}) = \sup_{f \in \mathcal{F}} \mathbb{E}[\inf\{n \in \mathbb{N}; X^*(f) = \{ES(x_1, \sigma_1, f), \ldots, ES(x_n, \sigma_n, f)\}\}].$$
Then, $\#RS(\mathcal{F}) \leq \mathbb{E}[\inf\{n \in \mathbb{N}; Disp((x_i)_{i \in \mathbb{N}}, n) \leq \epsilon\}].$

Proof: By Eq. 1, $X^*(f) \subset \{ES(x_1, \sigma_1, f), \ldots, ES(x_n, \sigma_n, f)\}$ for all $n \in \mathbb{N}$. On the other hand, if $Disp((x_i)_{i \in \mathbb{N}}, n) < \epsilon$, then Eq. 3 implies

$$\#RS(\mathcal{F}) \leq \sup_{f \in \mathcal{F}} \mathbb{E}[\inf\{n \in \mathbb{N}; X^*(f) \subset B(x_1, \epsilon) \cup B(x_2, \epsilon) \cup \cdots \cup B(x_n, \epsilon)\}]$$

and this is at most n; hence the expected result.   □

The bound is tight for some simple cases, e.g., optima distributed on a grid with sphere-like functions on Voronoi-cells built on the optima, and ES converging locally:

$$X^*(f) = \{(k_1\epsilon, k_2\epsilon, \ldots, k_D\epsilon); (k_1, \ldots, k_D) \in [[0, \lfloor 1\epsilon \rfloor]]^D\}. \tag{4}$$

The dispersion $Disp$ can therefore be used for quantifying the quality of a sequence of restart points. The optimal dispersion is reached by some grid-based sampling (e.g. Sukharev grids [16]) reaching $Disp(x, n) = O(n^{1/D})$. The **dispersion complexity**, i.e., the number n of points required for reaching dispersion ϵ is then[3]:

$O\left((1/\epsilon)^D \log(1/\epsilon)\right)$ for all low-discrepancy sequences; th. 6.6 p152,   (C1)
$O\left((1/\epsilon)^D\right)$ for Halton or Hammersley (Section 2); th. 6.12+6.13 p157,   (C2)
$\Omega\left((1/\epsilon)^D\right)$ for all sequences or point sets; th. 6.8 p154,   (C3)
$\Omega\left((1/\epsilon)^D \log(1/\epsilon)\right)$ for random sequences (expected value; see Theorem 2)(C4)

The results above are based on the notion of discrepancy. In particular, low-discrepancy (also termed quasi-random) sequences have *extreme discrepancy* $O(\log(n)^D/n)$. However, only complexity result (C1) will be necessary here.

To the best of our knowledge, complexity (C4) for random sequences has not yet been published. It can nonetheless easily be derived by reduction to the classical Coupon collector theorem (proof omitted for space reasons):

Theorem 2 (Dispersion of random points). *Consider* $x_1, \ldots, x_n, \ldots$ *randomly independently uniformly distributed in* $[0, 1]^D$ *and* $\epsilon > 0$. *Then* $\mathbb{E}[\inf\{n \in \mathbb{N}; Disp(x, n) < \epsilon\}] = \theta\left(1/\epsilon^D \log(1/\epsilon)\right)$.

From Proposition 1, and the complexity of dispersions above, it comes

Corollary 1 (Complexity in terms of number of restarts). *Let* $(\mathcal{F}^{(k)})_{k \in \mathbb{N}}$ *be a family of sets of fitness functions defined on* $[0, 1]^D$ *for some* D. *Suppose that for each* $k \in \mathbb{N}$, $\mathcal{F}^{(k)}$ *has strong local property (Eq. 3) with values* $\epsilon^{(k)}, \sigma_0^{(k)}$. *Then, for some* $C > 0$, *a quasi-random restart with Halton sequence[4] and* $\sigma_i < \sigma_0$ *ensures that*

$$(\forall k \in \mathbb{N}), (\forall f \in \mathcal{F}^{(k)}), (X^*(f) \subset \{ES(x_i, \sigma_i, f); i \in [[1, C/\epsilon^D]]\}). \tag{5}$$

This is not true for random restart, and C/ϵ^D *cannot be replaced by* $o(1/\epsilon^D)$.

---

[3] All references are to be found in [19]. See references therein for more details.
[4] Or any sequence with dispersion complexity (C1).

Algorithm 1. This generic algorithm includes Random restart with Decreasing Step-size (RDS) as well as Quasi-Random restart with Decreasing Step-size (QRDS) algorithm. Constant, linearly or quadratically decreasing versions can be implemented on line 5. The case with murder is the case $d > 0$ (line 14).

Require: $(x_1, x_2, \ldots)$ sequence of starting points, (d, σ^*) precision thresholds
1: $n = 0, optimaFound = \emptyset$
2: **while** Maximum number of evaluations not reached, and all optima not found **do**
3:      $n \leftarrow n + 1$
4:      $y = x_n$ // n^{th} point the sequence of starting points
5:      $\sigma = nextSigma(n, \sigma_0)$ // constant or decreasing σ
6:      $State \leftarrow alive$
7:      **while** $State = alive$ **do**
8:          $y' = y + \sigma \mathcal{N}(0, I_d)$ // $\mathcal{N}(0, I_d)$ is an isotropic Gaussian random variable
9:          **if** y' better than y **then**
10:             $y = y'$; $\sigma = 2\sigma$
11:         **else**
12:             $y = y'$; $\sigma = 2^{-1/4}\sigma$
13:         **for all** $(opt \in optimaFound)$ **do**
14:             **if** $(||y - opt|| < d)$ or $(\sigma < \sigma^*)$ **then**
15:                 $State \leftarrow dead$
16:     $optimaFound = optimaFound \cup \{y\}$

Proof: Eq. 5 is a consequence of (C1) and Proposition 1.

The fact that this is not true for random restart is the application of complexity (C4) to the particular case shown above (Eq. 4).

Finally, $\Omega(1/\epsilon^D)$ restarts are needed in order to find the $\Omega(1/\epsilon^D)$ optima in Eq. 4: it is hence not possible in the general case to replace C/ϵ^D by $o(1/\epsilon^D)$. $\square$

4   Experimental Results

This section presents comparative experimental results for some particular instances of restart algorithms to which the theoretical results of above Section 3 can be applied. The main ingredient of a restart algorithm is its embedded 'local' optimizer. The choice made here is that of an Evolution Strategy, for the robustness of this class of algorithm. However, only local convergence properties are required, far from any sophisticated variant designed for multimodal fitness functions for instance [26,9,1]. Furthermore, the testbed we want to compare to [22] does not require large values of λ, nor covariance matrix adaptation. Hence the simple (1+1)-ES with $\frac{1}{5}^{th}$-rule was chosen, as it gives very good results in very short time according to some preliminary runs. Finally, a (1+1)-ES satisfies the hypotheses of Theorem 1, and hence all results of Section 3 apply depending only on the properties of the sequence of starting points and initial step-sizes.

The precise instances of restart algorithm under scrutiny here are defined in Algorithm 1: this includes random and quasi-random sequences of starting points in the domain given as input (line 4), as well as different strategies for the step-size change from one run to the other, depending on line 5: function $nextSigma$ can return a constant value σ_0, a linearly decreasing value $(\sigma_0/(n+1))$ or a quadratically decreasing value $(\sigma_0/(n+1)^2)$.

One run of the local algorithm can be killed when either the step-size goes beyond a given thresholds σ^* (defaulted to 10^{-6} unless otherwise stated), or when the current solution gets too close to a previously discovered local optimum, up to a tolerance d (line 14). In the latter case, the algorithm is said *with murder*. But the choice of the threshold distance d is not trivial: a too small d wastes time, and too large values give imprecise results. Parameter d is a clear and strong drawback of the proposed algorithms, as the results are very sensitive to the value of d. Note however that the same remark holds for the modified clearing approach which outperformed by far all other methods in [22]. Nevertheless, because of this parameter, the generality of the results shown below is limited. In particular, we will limit our claims to the comparison with the results in [22], obtaining results that are comparable with the very good results of the modified clearing, with a similar parameter d, but with a simple restart algorithm.

Let us define K as the number of optima for the murder operator (the murder operator is not required for those complexity bounds). Let us first compare the complexity of the proposed restart algorithm with those provided in [22] (Table 1 (left)). The complexity for RDS/QRDS is immediate. In the case of murder operator, the complexity bound holds provided that K and d are such that the proportion of the domain which is forbidden by the murder operator is never larger than a fixed proportion of the whole domain. λ is the population size, 1 in our case (but the complexity results hold for any value of λ). RTS [10,11] and SCGA [15] stand for Restricted Tournament Selection and Species Conserving Genetic Algorithm respectively. w is window size of RTS. N_c is the number of clusters.

Let us then test the different approaches on the test functions provided in [22]. The first function is a n-dimensional sine, with 5^n peaks, defined on $[0,1]$ as $f(x) = 1 - \frac{1}{n}\sum_{i=1}^{n}(1 - \sin^6(5\pi x_i))$. In all experiments with this function, a point x^* will be considered an optimum if $f(x^*) > 0.997$.

Table 1. Left: Complexity of various algorithms. Most results are from [22] (see text). Right: Results on the sine function in dimension 1, for RDS and WRDS, and different strategies for σ. The last column is the percentage of additional evaluations when using random instead of quasi-random.

Algorithm	Complexity
RDS/QRDS, Clearing	$\Theta(\lambda)$
Deterministic or probabilistic crowding	$\Theta(\lambda)$
RTS	λw
QRDS/RDS with murder	$O(\lambda K)$
SCGA	$\Omega(\lambda)$ and $O(\lambda^2)$
Sharing, modified clearing	$\Theta(\lambda^2)$
Clustering	$\Theta(\lambda N_c)$

	Number of evaluations	Number of restarts	Computational overhead for RDS vs QRDS
Constant initial step-size $\sigma_0 = 0.1$			
RDS	1652	13.7	1%
QRDS	1628	13.41	
Constant initial step-size $\sigma_0 = 0.01$			
RDS	740 ± 15.81	6.07 ± 0.23	64%
QRDS	452 ± 8.95	2.3 ± 0.11	
Constant initial step-size $\sigma_0 = 0.001$			
RDS	808 ± 18.37	6.44 ± 0.25	79%
QRDS	451 ± 8.54	2.21 ± 0.10	
Quadratically decreasing step-size, $\sigma_0 = 1.$			
RDS	726 ± 18.34	6.73 ± 0.28	46%
QRDS	498 ± 10.15	3.72 ± 0.15	
Quadratically decreasing step-size, $\sigma_0 = 0.1$			
RDS	755 ± 17.40	6.65 ± 0.25	64%
QRDS	461 ± 8.74	2.79 ± 0.11	

Another important test function is the hump function defined in [22] as: $f(x) = h \max\left(1 - (\inf_k ||x - x_k||/r)^{\alpha_k}, 0\right)$ where $\alpha_k = 1$, $r = 1.45$, $h = 1$; sequence $x_1, \ldots, x_k$ is randomly drawn in domain $[0,1]^D$; the problem is used in dimension 25 with $k = 50$, the most difficult case according to [22].

4.1   Constant Initial Step-Size Is Dangerous

First introduced in [23], the idea of decreasing σ at each restart is somewhat natural, when considering the convergence proof (see Remark 1). From the results on the sine function in dimension 1 (Table 1-right), it is clear that a too large fixed σ leads to poor results, while a too small σ also hinders the performances. Indeed, when looking for all optima, a large initial step-size might be helpful in order to avoid poor local optima. However it is a bad idea for finding optima close to the frontier of the domain when there are big basins of attractions. Furthermore, the proved version, with quasi-random restarts and decreasing σ, performs well without any tuning, though less efficiently than the version with a *posteriori* chosen fixed σ. Yet, its good performances independently of any parameter tuning is a strong argument for the proved method.

4.2   Validating Quasi-Random Sequences

Let us now focus on the comparison between the use of quasi-random vs random restarts, i.e., RDS vs QRDS. From results on the multi-dimensional sine function (Tables 1-right and 2), it is clear that quasi-random becomes more and more efficient as the number of optima increases.

Table 2. RDS vs QRDS on the multidimensional sine function: number of evaluations (number of restarts), averaged over 30 runs, for finding the 5^D optima in dimension D. Here the step-size is quadratically decreasing with initial step-size $\sigma_0 = 0.1$.

Dimension D	RDS	QRDS	Additional cost for RDS over QRDS
3	143986 (803)	109128 (609)	32%
2	11673 (98)	8512 (71)	37%
1	777 (13)	447 (8)	74%

Let us now revisit the sine function in dimension 1, and increase the number of optima by increasing its parameter K from 5 to 50, 500, and 1000. Other parameters of the algorithms are here $\sigma_{init} = 10^{-1}/K$, $\sigma^* = 5.10^{-4}/K$, and $d = 0.5/K$ for the murder threshold. The performances reported in Table 3-left witness the computational effort until all optima are found, i.e. there is a point of fitness > 0.997 at distance lower than half the distance between optima (please remember than 0.997 is the chosen threshold in this benchmark). Those results show that QRDS becomes more and more efficient when compared to RDS as the number of optima increases. Furthermore, the murder operator is clearly highly efficient.

Table 3. Left: Performance of RDS, QRDS, and QRDS with Murder when the number of optima on the 1-D sine function increases. The last column shows the percentage of additional evaluations for RDS over the given algorithm. Right: Comparative results on the 25-dimensional hump problem. All results but RDS/QRDS are taken from [22], where the modified clearing is reported to have found all 50 optima in all of the 30 runs. RDS and QRDS being much faster (see Table 1), 100 experiments have been run easily, and almost all 50 optima have been consistently found. See text for details.

	Nb of evaluations	Nb of restarts	Ratio
	5 optima		
RDS	1484 ± 33.59	5.79 ± 0.24	25 %/
QRDS	1187 ± 30.27	4.17 ± 0.20	152%
QRDS+M	588 ± 3.24	1.31 ± 0.04	
	50 optima		
RDS	30588 ± 156.28	171.45 ± 1.12	46% /
QRDS	20939 ± 100.04	102.24 ± 0.71	364%
QRDS+M	6583 ± 2.39	17.39 ± 0.05	
	500 optima		
RDS	470080 ± 548.60	2900.75± 3.96	67% /
QRDS	281877± 279.15	1540.5± 2.01	603%
QRDS+M	66789 ± 3.07	161.92 ± 0.07	
	1000 optima		
RDS	1009144 ± 747.05	6298.16 ± 5.40	61 % /
QRDS	627696 ± 409.41	3545.61 ± 2.96	655%
QRDS+M	133587 ± 3.11	320.8 ± 0.07	

Algorithm	Nb of optima found
Sharing	$\simeq 0$
D./P. Crowding	$\simeq 0$ / 0
RTS	$\simeq 0$
SCGA	$\simeq 0$
Clustering	$\simeq 0$
Clearing	$\simeq 43$
Modified Clearing	$\simeq 50$
RDS	49.92 (over 100 runs)
QRDS	49.95 (over 100 runs)

4.3 Comparison with Modified Clearing

In this section, we compare the restart as previously defined (quasi-random restarts, decreasing σ and murder) to the best techniques in the survey [22]. Interestingly, some algorithms tested in [22] (Probabilistic Crowding and SCGA) could not even find all optima of the simple sine function in dimension 1. Note that QRDS finds all optima within a few hundred evaluations, whereas according to [22] nearly 100 generations of population size 50 are necessary for finding the 5 optima for all methods. However, because [22] does not provide quantitative results, precise comparisons are not possible here. Therefore, the following comparative results use the hump function, for which [22] provides extensive experiments with detailed results. This function is particularly challenging, because [22] points out that no algorithm except their modified clearing can solve the problem. The modified clearing, however, finds all optima in each of the 30 runs, whereas all methods (as reported in Table 1), except clearing, do not even find a single optimum. Table 3-right reports the results of RDS and QRDS for the hardest instance, in dimension 25 where the number of optima is 50. Averaged over 100 runs, both methods find almost all 50 optima (on average more than 49.9).

5 Conclusion

This paper has introduced some generic framework for restart algorithms embedding a (local) optimization algorithm. With limited hypotheses about the local properties of the embedded algorithm, in particular with respect to some initial parameter σ describing the size of its basins of attraction, we have proved some convergence properties with speed depending on the dispersion of the sequence of starting points, independently of any other parameter of the embedded algorithm.

Actual instances of the generic algorithm have then been proposed, embedding a (1+1)-Evolution Strategy with $\frac{1}{5}^{th}$ rule, for which the initial step-size plays the role of parameter σ above. Random and quasi-random sequences of starting points can be used. The proposed algorithms have mathematically proved convergence properties. Thanks to the decreasing of the initial step-size to 0, the convergence is proved independently of initialization parameters. Furthermore, experimental validation of the proposed algorithm have been conducted, comparing random and quasi-random sequences of starting points, and judging performance with respect to other previously published algorithms [22].

A first conclusion is the not surprising superiority of quasi-random restarts over random restarts. The improvement due to QR points is moderate (a logarithmic factor), but regular and increasing when considering more complicated cases (more optima to be found). We have no experimental evidence in this paper or in the literature for the superiority of any algorithm (whatever complicated and computationally expensive) over the simple restart algorithm with decreasing σ and quasi-random initializations that has been proposed here. QRDS performed equally or better than modified clearing [22], whilst keeping a linear computational cost. All methods with non quadratic cost benchmarked in [22] are much less efficient than QRDS, RDS or modified clearing, while RDS and QRDS are much cheaper than modified clearing.

The murder operator, that kills runs that get close to previously found optima, was found highly beneficial (up to more than 700 % speed-up for 500 optima). However, QRDS with murder operator has the same weakness than modified clearing: it introduces a crucial parameter taking into account the distance between different optima, somewhat similar to σ_{clear} parameter of modified clearing (that has additionally two arbitrary constants 1.5 and 3 for solving the difficult hump function).

Let us now discuss the limitations of our analysis, and some general elements. An important element in this work is the choice of benchmark functions. The hump function is in fact a distribution of fitness functions, and not only a fitness function, or random translations of a fitness function. We agree with the authors of [22] that solving the hump function is by no means an easy task. However, this benchmark has its limitations: as the function is locally very simple, and as a very loose precision is required, the best choice for the embedded ES is the 1+1-ES with a very loose halting condition: this is perhaps not the less realistic scenario, but this certainly does not covers all cases. A strong advantage of the hump function is that it cannot easily be overfitted. However, all successful methods on the hump function have a parameter which is intuitively related to the distance between optima: This suggests that, in spite of the random part in the hump function, some overfitting nevertheless happens when tuning an algorithm on the hump function. Also, we feel these experiments are definitely convincing regarding the importance of mathematical analysis in MMO: it is otherwise very easy to conclude positively about an algorithm, without seeing its precise limitations. Thanks to theoretical analysis, we could clearly see that decreasing the step-size provides a real advantage (a decreasing step-size provides consistency) and that quasi-randomizing the restarts provides an improvement - and to the best of our knowledge, as QRDS has the best experimental results and the best proved results, this paper provides a clear and simple baseline for future research in MMO. A case in which clearing approaches might perform better than the proposed approach is the parallel case. The murder operator might be less efficient if several populations evolve simultaneously. The proposed approach has been designed for the case of MMO in which the goal is to locate all optima, opposed to the case where the goal is to locate the global optimum, but there exist many local optima. However, these two forms of MMO are probably not so different - if the local optima have fitness value close to the one of the global optimum x^*, we have to check all local optima.

References

1. Augr, A., Hansen, N.: A restart CMA evolution strategy with increasing population size. In: Proceedings of the IEEE Congress on Evolutionary Computation, CEC 2005, pp. 1769–1776 (2005)
2. Auger, A., Jebalia, M., Teytaud, O.: Algorithms (X, sigma, eta): Quasi-random Mutations for Evolution Strategies. In: Talbi, E.-G., Liardet, P., Collet, P., Lutton, E., Schoenauer, M. (eds.) EA 2005. LNCS, vol. 3871, pp. 296–307. Springer, Heidelberg (2006)
3. Auger, A., Teytaud, O.: Continuous lunches are free plus the design of optimal optimization algorithms. Algorithmica 57(1), 121–146 (2010)
4. Beyer, H.-G., Sendhoff, B.: Covariance Matrix Adaptation Revisited – The CMSA Evolution Strategy –. In: Rudolph, G., Jansen, T., Lucas, S., Poloni, C., Beume, N. (eds.) PPSN X. LNCS, vol. 5199, pp. 123–132. Springer, Heidelberg (2008)
5. Beyer, H.-G.: The Theory of Evolution Strategies. Springer, Heidelberg (2001)
6. DeJong, K.: The Analysis of the Behavior of a Class of Genetic Adaptive Systems. PhD thesis, University of Michigan, Ann Harbor (1975); Dissertation Abstract International 36(10), 5140B (University Microfilms No 76-9381)
7. Georgieva, A., Jordanov, I.: A hybrid meta-heuristic for global optimisation using low-discrepancy sequences of points. Computers and Operations Research - Special Issue on Hybrid Metaheuristics (in press)
8. Goldberg, D.E., Richardson, J.: Genetic algorithms with sharing for multimodal-function optimization. In: Grefenstette, J.J. (ed.) ICGA, pp. 41–49. Lawrence Erlbaum Associates (1987)
9. Hansen, N., Kern, S.: Evaluating the CMA Evolution Strategy on Multimodal Test Functions. In: Yao, X., Burke, E.K., Lozano, J.A., Smith, J., Merelo-Guervós, J.J., Bullinaria, J.A., Rowe, J.E., Tiňo, P., Kabán, A., Schwefel, H.-P. (eds.) PPSN VIII. LNCS, vol. 3242, pp. 282–291. Springer, Heidelberg (2004)
10. Harik, G.: Finding multiple solutions in problems of bounded difficulty. Technical Report 94002, University of Illinois at Urbana Champaign (1994)
11. Harik, G.: Finding multimodal solutions using restricted tournament selection. In: Eshelman, L.J. (ed.) Proc. Sixth International Conference on Genetic Algorithms, pp. 24–31. Morgan Kaufmann (1995)
12. Jones, D.R., Schonlau, M., Welch, W.J.: Efficient global optimization of expensive black-box functions. J. of Global Optimization 13(4), 455–492 (1998)
13. Kimura, S., Matsumura, K.: Genetic algorithms using low-discrepancy sequences. In: GECCO, pp. 1341–1346 (2005)

14. L'Ecuyer, P., Lemieux, C.: Recent Advances in Randomized Quasi-Monte Carlo Methods, pp. 419–474. Kluwer Academic (2002)
15. Li, J., Balazs, M., Parks, G., Clarkson, P.: A species conserving genetic algorithm for multimodal function optimization. Evolutionary Computation 10(3), 207–234 (2002)
16. Lindemann, S.R., LaValle, S.M.: Incremental low-discrepancy lattice methods for motion planning. In: Proceedings IEEE International Conference on Robotics and Automation, pp. 2920–2927 (2003)
17. Mahfoud, S.W.: Niching methods for genetic algorithms. PhD thesis, University of Illinois at Urbana-Champaign, Champaign, IL, USA (1995)
18. Mengshoel, O.J., Goldberg, D.E.: Probabilistic crowding: Deterministic crowding with probabilisitic replacement. In: Banzhaf, W., Daida, J., Eiben, A.E., Garzon, M.H., Honavar, V., Jakiela, M., Smith, R.E. (eds.) Proceedings of the Genetic and Evolutionary Computation Conference, Orlando, Florida, USA, July 13-17, vol. 1, pp. 409–416. Morgan Kaufmann (1999)
19. Niederreiter, H.: Random Number Generation and Quasi-Monte Carlo Methods. SIAM, Philadelphia (1992)
20. Owen, A.B.: Quasi-Monte Carlo sampling. In: Jensen, H.W. (ed.) Monte Carlo Ray Tracing: Siggraph 2003 Course 44, pp. 69–88. SIGGRAPH (2003)
21. Pétrowski, A.: A clearing procedure as a niching method for genetic algorithms. In: International Conference on Evolutionary Computation, pp. 798–803 (1996)
22. Singh, G., Kalyanmoy Deb, D.: Comparison of multi-modal optimization algorithms based on evolutionary algorithms. In: GECCO 2006: Proceedings of the 8th Annual Conference on Genetic and Evolutionary Computation, pp. 1305–1312. ACM, New York (2006)
23. Teytaud, F., Teytaud, O.: **Log**(λ) Modifications for Optimal Parallelism. In: Schaefer, R., Cotta, C., Kołodziej, J., Rudolph, G. (eds.) PPSN XI. LNCS, vol. 6238, pp. 254–263. Springer, Heidelberg (2010)
24. Teytaud, O.: When Does Quasi-random Work? In: Rudolph, G., Jansen, T., Lucas, S., Poloni, C., Beume, N. (eds.) PPSN X. LNCS, vol. 5199, pp. 325–336. Springer, Heidelberg (2008)
25. Teytaud, O., Gelly, S.: DCMA: yet another derandomization in covariance-matrix-adaptation. In: Thierens, D., et al. (eds.) ACM-GECCO 2007, pp. 955–963. ACM, New York (2007)
26. van den Bergh, F., Engelbrecht, A.: Cooperative learning in neural networks using particle swarm optimizers (2000)
27. Vandewoestyne, B., Cools, R.: Good permutations for deterministic scrambled halton sequences in terms of l2-discrepancy. Computational and Applied Mathematics 189(1,2), 341–361 (2006) bibitemiccama Vandewoestyne, B., Cools, R.: Good permutations for deterministic scrambled halton sequences in terms of l2-discrepancy. Computational and Applied Mathematics 189(1,2), 341–361 (2006)
28. Villemonteix, J., Vazquez, E., Walter, E.: An informational approach to the global optimization of expensive-to-evaluate functions. Journal of Global Optimization 44(4), 509–534 (2009)
29. Yin, X., Germay, N.: A fast genetic algorithm with sharing scheme using cluster analysis methods in multimodal function optimization. In: Albrecht, R.F., Steele, N.C., Reeves, C.R. (eds.) Artificial Neural Nets and Genetic Algorithms, pp. 450–457. Springer (1993)

Local Optima Networks with Escape Edges

Sébastien Vérel[2], Fabio Daolio[3], Gabriela Ochoa[1], and Marco Tomassini[3]

[1] Department of Computing Science and Mathematics, University of Stirling, Stirling, Scotland
[2] INRIA Lille - Nord Europe and University of Nice Sophia-Antipolis, France
[3] Faculty of Business and Economics, University of Lausanne, Lausanne, Switzerland

Abstract. This paper proposes an alternative definition of edges (*escape edges*) for the recently introduced network-based model of combinatorial landscapes: *Local Optima Networks (LON)*. The model compresses the information given by the whole search space into a smaller mathematical object that is the graph having as vertices the local optima and as edges the possible weighted transitions between them. The original definition of edges accounted for the notion of transitions between the basins of attraction of local optima. This definition, although informative, produced densely connected networks and required the exhaustive sampling of the basins of attraction. The alternative escape edges proposed here do not require a full computation of the basins. Instead, they account for the chances of escaping a local optima after a controlled mutation (e.g. 1 or 2 bit-flips) followed by hill-climbing. A statistical analysis comparing the two LON models for a set of NK landscapes, is presented and discussed. Moreover, a preliminary study is presented, which aims at validating the LON models as a tool for analyzing the dynamics of stochastic local search in combinatorial optimization.

1 Introduction

The performance of heuristic search algorithms crucially depends on the structural aspects of the spaces being searched. An improved understanding of this dependency, can facilitate the design and further successful application of these methods to solve hard computational search problems. Local optima networks (LON) have been recently introduced as a novel model of combinatorial landscapes [10,11]. This model allows the use of complex network analysis techniques [7] in connection with the study of fitness landscapes and problem difficulty in combinatorial optimization. The model is based on the idea of compressing the information given by the whole problem configuration space into a smaller mathematical object, which is the graph having as vertices the optima configurations of the problem and as edges the possible transitions between these optima. This characterization of landscapes as networks has brought new insights into the global structure of the landscapes studied, particularly into the distribution of their local optima. Moreover, some network features have been found to correlate and suggest explanations for search difficulty on the studied domains.

The definition of the edges in the LON model critically impacts upon its descriptive power with regards to heuristic search. The initial definition of edges in [10,11], *basin-transition* edges, accounted for the notion of transitions between the local optima basins' frontiers. This definition, although informative, produces highly connected networks and requires the exhaustive sampling of the basins of attraction. We explore in

J.-K. Hao et al. (Eds.): EA 2011, LNCS 7401, pp. 49–60, 2012.

this article an alternative definition of edges, which we term *escape* edges, that does not require a full computation of the basins. Instead, the edges account for the chances (of a prospective heuristic search algorithm) of escaping a local optima after a controlled mutation (e.g. 1 or 2 bit-flips) followed by hill-climbing. This new definition produces less dense and easier to build LONs, which are more amenable to sampling and get us closer to a fitness landscape model that can be used to understand (and eventually exploit) the dynamics of local search on combinatorial problems.

The first goal of the present study is to compare and explore the relationships between the two LON models, based on (i) basin-transition edges and (ii) escape edges, respectively. Thereafter, we present a preliminary study that aims at validating the LON models in their descriptive power of the dynamics of stochastic local search algorithms. We conduct this validation by considering the behavior of two well-known stochastic local search heuristics, namely, Tabu Search [4] and Iterated Local Search [6]. The well known family of NK landscapes [5] is used in our study.

The article is structured as follows. Section 2, includes the relevant definitions and algorithms for extracting the LONs. Section 3, describes the experimental design, and reports a comparative analysis of the extracted networks of the two models. Section 4, presents our model validation study. Finally, section 5 discusses our main findings and suggest directions for future work.

2 Definitions and Algorithms

A Fitness landscape [9] is a triplet (S, V, f) where S is a set of potential solutions i.e. a search space, $V : S \longrightarrow 2^S$, a neighborhood structure, is a function that assigns to every $s \in S$ a set of neighbors $V(s)$, and $f : S \longrightarrow R$ is a fitness function that can be pictured as the *height* of the corresponding solutions. In our study, the search space is composed of binary strings of length N, therefore its size is 2^N. The neighborhood is defined by the minimum possible move on a binary search space, that is the single bit-flip operation. Thus, for a bit string s of length N, the neighborhood size is $|V(s)| = N$.

The $HillClimbing$ algorithm to determine the local optima and therefore define the basins of attraction, is given in Algorithm 1. It defines a mapping from the search space S to the set of locally optimal solutions S^*. Hill climbing algorithms differ in their so-called *pivot-rule*. In best-improvement local search, the entire neighborhood is explored and the best solution is returned, whereas in first-improvement, a neighbor is selected uniformly at random and is accepted if it improves on the current fitness value. We consider here a best-improvement local searcher (see Algorithm 1). For a comparison between first and best-improvement LON models, the reader is referred to [8]

2.1 Nodes

As discussed above, a best-improvement local search algorithm based on the 1-move operation is used to determine the local optima. A local optimum (LO), which is taken to be a maximum here, is a solution s^* such that $\forall s \in V(s), f(s) \leq f(s^*)$.

Let us denote by $h(s)$, the stochastic operator that associates to each solution s, the solution obtained after applying the best-improvement hill-climbing algorithm (see

---

Algorithm 1. Best-improvement local search (hill-climbing)

Choose initial solution $s \in S$
repeat
    choose $s' \in V(s)$, such that $f(s') = max_{x \in V(s)} f(x)$
    if $f(s) < f(s')$ **then**
        $s \leftarrow s'$
    end if
until s is a Local optimum

Algorithm 1) until convergence to a LO. The size of the landscape is finite, so we can denote by LO_1, LO_2, $LO_3 \ldots, LO_p$, the local optima. These LOs are the vertices of the *local optima network*.

2.2 Basin-Transition Edges

The basin of attraction of a local optimum $LO_i \in S$ is the set $b_i = \{s \in S \mid h(s) = LO_i\}$. The size of the basin of attraction of a local optimum i is the cardinality of b_i, denoted $\sharp b_i$. Notice that for non-neutral[1] fitness landscapes, as are standard NK landscapes, the basins of attraction as defined above, produce a partition of the configuration space S. Therefore, $S = \cup_{i \in S^*} b_i$ and $\forall i \in S \, \forall j \neq i, \, b_i \cap b_j = \emptyset$.

We can now define the weight of an edge that connects two feasible solutions in the fitness landscape. For each pair of solutions s and s', $p(s \rightarrow s')$ is the probability to pass from s to s' with the given neighborhood structure. In the case of binary strings of size N, and the neighborhood defined by the single bit-flip operation, there are N neighbors for each solution, therefore, considering a uniform selection of random moves:
if $s' \in V(s)$, $p(s \rightarrow s') = \frac{1}{N}$ and
if $s' \notin V(s)$, $p(s \rightarrow s') = 0$.
The probability to go from solution $s \in S$ to a solution belonging to the basin b_j, is[2]:

$$p(s \rightarrow b_j) = \sum_{s' \in b_j} p(s \rightarrow s') \quad .$$

Thus, the total probability of going from basin b_i to basin b_j, i.e. the weight w_{ij} of edge e_{ij}, is the average over all $s \in b_i$ of the transition probabilities to solutions $s' \in b_j$:

$$p(b_i \rightarrow b_j) = \frac{1}{\sharp b_i} \sum_{s \in b_i} p(s \rightarrow b_j) \quad .$$

2.3 Escape Edges

The escape edges are defined according to a distance function d (minimal number of moves between two solutions), and a positive integer $D > 0$.

---

[1] For a definition of basins that deals with neutrality, the reader is referred to [11].
[2] Notice that $p(s \rightarrow b_j) \leq 1$ and notice also that this definition, disregarding the fitness values, is purely topological and is not related to any particular search heuristic.

There exists an edge e_{ij} between LO_i and LO_j if it exists a solution s such that $d(s, LO_i) \leq D$ and $h(s) = LO_j$. The weight w_{ij} of this edge is then: $w_{ij} = \sharp\{s \in S \mid d(s, LO_i) \leq D$ and $h(s) = LO_j\}$, which can be normalized by the number of solutions within reach w.r.t. such a distance $\sharp\{s \in S \mid d(s, LO_i) \leq D\}$.

2.4 Local Optima Network

The weighted local optima network $G_w = (N, E)$ is the graph where the nodes $n_i \in N$ are the local optima, and there is an edge $e_{ij} \in E$, with weight $w_{ij} = p(b_i \rightarrow b_j)$, between two nodes n_i and n_j if $p(b_i \rightarrow b_j) > 0$.

According to both definitions of edge weights, $w_{ij} = p(b_i \rightarrow b_j)$ may be different than $w_{ji} = p(b_j \rightarrow b_i)$. Thus, two weights are needed in general, and we have an oriented transition graph.

Figure 1 depicts a representative example of the alternative LON models. The figures corresponds to a real NK landscape with N = 18, K = 2, which is the lowest ruggedness value explored in our study. The left plot illustrates the basin-transition edges, while the center and right plots the escape edges with $D = 1$ and $D = 2$, respectively. Notice that the basin-transition edges (left) produce a densely connected network, while the escape edges produce more sparse networks.

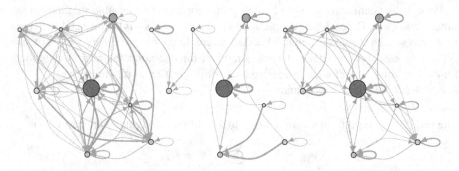

Fig. 1. Local optima network of an NK-landscape instance with N = 18, K = 2. Left: basin-transition edges. Center and Right: escape edges with $D = 1$ and $D = 2$, respectively. The size of the circles is proportional to the logarithm of the size of the corresponding basins of attraction; the darker the color, the better the local optimum fitness. The edges' width scales with the transition probability (weight) between local optima, according to the respective definitions. Notice that the basin-transition edges model (Left) is much more densely connected.

3   Comparative Analysis of the LON Models

In this section, we compare the LONs resulting from the different edges definitions discussed above. We chose to perform this analysis on the NK-model artificial landscapes, primarily to be able to compare directly with previous work [10,11], but also because this problem provides a framework that is of general interest in studying the structure of complex combinatorial problems [5].

The NK family of correlated landscapes is in fact a problem-independent model for constructing multimodal landscapes that can gradually be tuned from smooth to rugged. In the model, N refers to the number of (binary) genes in the genotype, i.e. the string length, and K to the epistatic interaction, i.e. the number of genes that influence a particular gene. By increasing the value of K from 0 to $N-1$, the landscapes can be tuned from smooth to rugged. The K variables that form the context of the fitness contribution of a gene, can be chosen according to different models, the two most widely studied being the *random neighborhood* model and the *adjacent neighborhood* model. As no significant differences between the two were found, neither in terms of the landscape global properties [5] nor in terms of their local optima networks (preliminary studies), we conduct our full study on the more general random model.

In order to minimize the influence of the random creation of landscapes, we considered 30 different and independent problem instances for each combination of N and K parameter values. In all cases, the measures reported are the average of these 30 landscapes. In the present study, $N = 18$ and $K \in \{2, 4, 6, 8, 10, 12, 14, 16, 17\}$, which are the largest possible parameter combinations that allow the exhaustive extraction of local optima networks. LONs for the two definitions of edges: (i) basin-transition and (ii) escape edges with $D \in \{1, 2\}$, were extracted and analyzed[3].

Table 1. Local optima network features. Values are averages over 30 random instances, standard deviations are shown as subscripts. K = epistasis value of the corresponding NK-landscape ($N = 18$); N_v = number of vertices; D_{edge} = density of edges ($N_e/(N_v)^2 \times 100\%$); L_{opt} = average shortest path to reach the global optimum ($d_{ij} = 1/w_{ij}$).

K	N_v	D_{edge} (%)			L_{opt}		
	all	Basin-trans.	Esc.$D1$	Esc.$D2$	Basin-trans.	Esc.$D1$	Esc.$D2$
2	$43.0_{27.7}$	$74.182_{13.128}$	$8.298_{4.716}$	$22.750_{9.301}$	$21.2_{8.0}$	$16.8_{4.7}$	$33.5_{14.1}$
4	$220.6_{39.1}$	$54.061_{4.413}$	$1.463_{0.231}$	$7.066_{0.810}$	$41.7_{10.5}$	$19.2_{5.1}$	$53.7_{12.4}$
6	$748.4_{70.2}$	$26.343_{1.963}$	$0.469_{0.047}$	$3.466_{0.279}$	$80.0_{19.1}$	$22.2_{3.9}$	$66.7_{12.9}$
8	$1668.8_{73.5}$	$12.709_{0.512}$	$0.228_{0.009}$	$2.201_{0.066}$	$110.1_{13.8}$	$24.0_{4.9}$	$76.6_{9.1}$
10	$3147.6_{109.9}$	$6.269_{0.244}$	$0.132_{0.004}$	$1.531_{0.036}$	$152.8_{19.3}$	$27.3_{5.0}$	$90.7_{8.4}$
12	$5270.3_{103.9}$	$3.240_{0.079}$	$0.088_{0.001}$	$1.115_{0.015}$	$185.1_{23.8}$	$30.3_{6.7}$	$108.3_{12.3}$
14	$8099.6_{121.1}$	$1.774_{0.035}$	$0.064_{0.001}$	$0.838_{0.009}$	$200.2_{16.0}$	$38.9_{9.6}$	$124.7_{8.6}$
16	$11688.1_{101.3}$	$1.030_{0.013}$	$0.051_{0.000}$	$0.647_{0.004}$	$211.8_{15.0}$	$47.9_{11.4}$	$146.2_{11.2}$
17	$13801.0_{74.1}$	$0.801_{0.007}$	$0.047_{0.000}$	$0.574_{0.002}$	$214.3_{17.5}$	$55.7_{12.5}$	$155.9_{12.2}$

3.1 Network Features and Connectivity

Number of Nodes and Edges. The 2nd column of Table 1, reports the number of nodes (local optima), N_v, which is the same for all the studied landscapes and models. The number of nodes increase exponentially with increasing values of K. The networks, however, have a different number of edges, as can be appreciated in the 3rd, 4th, and

---

[3] Some of the the tools for fitness landscape analysis and the local search heuristics, were used from the "ParadisEO" library [2]; data treatment and network analysis are done in "R" with the "igraph" package [3].

5th columns of Table 1, which report the number of edges normalized by the square of the number of nodes (density of edges). Clearly, the density is higher for the basin-transition edges, followed by the escape edges with $D = 2$, and the smaller density corresponds to $D = 1$. The trend is, however, that density decreases steadily with increasing values of K, which supports the correlation between the two models.

With the basin-transition edges, LONs are densely connected, especially when K is low: 74% and 54% of all possible edges are present, on average, for $K \in \{2, 4\}$. The escape edges produce sparsely connected graphs. Indeed, the $D = 1$-escape edge model, produces networks that are not completely connected, with the number of connected components ranging between 1.67 and 8.37 in average. The global optimum, though, always happens to belong to the largest connected component, which, in our analysis, comprises an average proportion of solutions raising, with increasing values of K, from 0.9392764 to 0.9999879.

The networks with escape edges and $D = 2$, are always connected. The density decreases with the epistasis degree. For high Ks, the density values are close to those of the basin-transition networks. Figure 2 (Left) illustrates what is happening in terms of the average degree of the outgoing links. First, notice that the difference with the basin-transition networks is maximal when K is between 4 and 12. Whereas for $D = 1$ the outgoing degree only increases from 1.7 to 5.5 across the range of K values, for $D = 2$ the growth is faster and reaches 78.2, not far from the 109.5 score of the LON with basin-transition edges. The size of the basins could provide an explanation for this: at high values of K basins are so small that a 2-bit mutation from the local optimum is almost enough to recover the complete topology.

Clustering Coefficient. LONs with basins-transitions edges present a somewhat symmetric structure: when two nodes i and j are connected, both edges e_{ij} and e_{ji} are present (even though their weights are in general different, $w_{ij} \neq w_{ji}$). Moreover, those connections often form triangular closures whose frequency is given by the global clustering coefficient. As Figure 2 (Right) shows, this measure of transitivity is lower with the escape-transition edges, but the difference could be due to the different number of

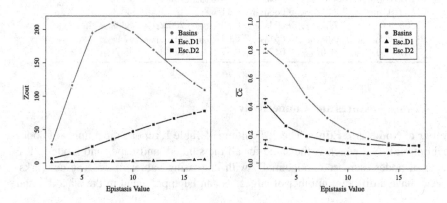

Fig. 2. Average out-degree (Left) and average clustering coefficient (Right) vs epistasis value

edges of those LONs. Values for $D = 1$ are remarkably low, even if the calculation disregarded the direction of edges. Overall though, the decreasing trend w.r.t. the landscape ruggedness, remains common. In other words, even with escape-transition edges, the clustering coefficient can be retained as a measure related to problem complexity: it decreases with the non-linearity of the (NK-) problem.

Shortest Paths. Due to the differences in topology, in the escape edges networks not all the paths are possible: few nodes might be disconnected, or they might not be reachable due to direction constraints (these are more "asymmetric" networks, as can be seen in Fig. 1). Thus, while evaluating shortest paths, only paths connecting reachable couples of nodes are averaged. Moreover, there are different ranges of weights, so the values displayed in Figure 3 have been normalized. An unexpected behavior can be observed for $D = 2$: the average path length peaks at $K = 4$ and stays always high. Maybe the increasing connectivity of nodes (see Fig. 2) counteracts their increase in numbers. However, some paths are more important than others, for example those who lead to the global optimum (see Fig. 3 (right)). With respect to these paths, all the LON models show the same trend: the paths increase in length as ruggedness increases.

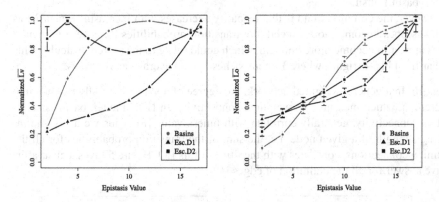

Fig. 3. Shortest paths over the LON vs epistasis value. Left: average geodesic distance between optima. Right: average shortest path to the global optimum. Each curve has been divided by its respective maximum value.

3.2 Characterization of Weights

Disparity. Figure 2 (Left) gave the average connectivity of vertices, counting outgoing links. One might then ask whether or not there are preferential directions when leaving a particular basin, i.e. if for a given node i, the weights w_{ij} (with $j \neq i$) are equivalent. For this purpose, we measure the disparity Y_2 [1], which gauges the heterogeneity of the contributions of the edges of node i to the total weight. For a large enough degree z_i, when there is not a dominant weight, then $Y_2 \approx 1/z_i$. The connectivity of LONs with escape edges with $D = 1$ is weak, and it is difficult to draw conclusions based on disparity only, as Figure 4 (left) illustrates. In the example illustrated, where $K = 4$, (i.e. relatively low epistasis, and so not a random structure), Y_2 approaches $1/z$ for

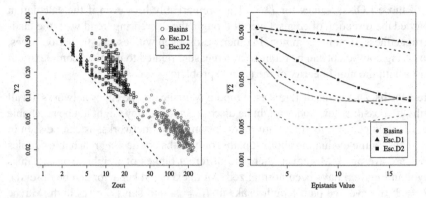

Fig. 4. Disparity of edges' weights. Left: scatter plot of disparity against outgoing degree (lin-log scale) for an instance with epistasis $K = 4$. Right: average vertex disparity for outgoing edges vs epistasis value. Dotted lines represent the inverse of the outgoing degree.

escape-$D = 1$, whereas it has distinctively higher values for both the escape-$D = 2$, and the basin-transition edges.

However, the common trend is that disparity decreases with increasing epistasis: as the landscapes become more rugged, the transition probabilities to leave a particular basins appear to become more uniform, which could relate to the search difficulty. This is clear from Fig 4 (right), where Y_2 approaches $1/z$ on average as K grows.

Strength. In a general weighted network, the degree of a vertex naturally extends into its strength, which measures its weighted connectivity. In LON model, basins size, as well as connectivity, generally correlate with fitness value [8]. Thus we ask if the incoming strength of a given node, i.e. the sum of the transition probabilities for all the incoming connections, correlates with the fitness of its LO. Figure 5 gives a clear affirmative answer for all the definition of edges.

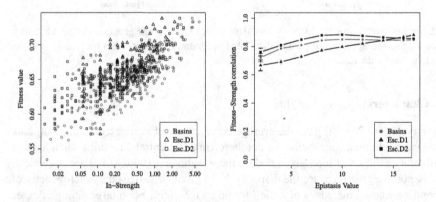

Fig. 5. Correlation between the strength of a node and its fitness value. Left: scatter plot of fitness against weighted connectivity (lin-log scale) for an instance with epistasis $K = 4$. Right: average correlation Spearman's coefficient between in-strength and fitness value vs epistasis value.

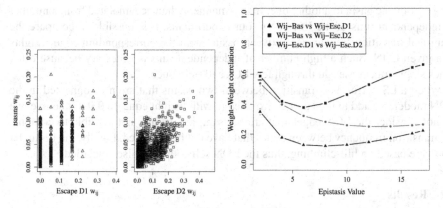

Fig. 6. Correlation between the weights resulting from the the new and the old definition of edges. Left: scatter plot for an instance with epistasis $K = 4$. Right: average Spearman's rank correlation coefficient vs epistasis value. No self-loops.

Correlation of Weights among Edges' Definitions. The LON models resulting from the alternative definition of edges show different structures but common trends w.r.t. the features that are related to problem difficulty. We conclude this subsection by directly comparing the transition probabilities that result from the alternative edge definitions. For this purpose, Figure 6 shows the Spearman's rank correlation between the corresponding rows of the weighted adjacency matrix, for different LON models of the same instance. The statistic is always positive, but, given the sparser nature of the escape-$D = .1$ networks, the result is weaker in this case. With $D = 2$, though, the correlation with the basin-transition definition is consistently good, which agrees with the previous findings on this topology. Indeed, as the landscape ruggedness increases, the correlation between $D = 1$ and $D = 2$ becomes smaller.

4 Model Validation: Local Search Dynamics

In this section, we analyze the connection between the LON model and dynamics of local search (LS) in an attempt to validate its descriptive power. Does a LS follow the edges of the LON? Which edge definition is the most accurate to predict the dynamics of a local search heuristic? This is a preliminary study on one particular NK-landscape instance with $N = 18$ and $K = 4$, a larger analysis will be the subject of future work.

4.1 Experiment Setup

We choose two simple but efficient stochastic LS heuristics, namely Iterated Local Search (ILS) [6], and Tabu Search (TS) [4]. Both are given $2.6 * 10^4$ function evaluations ($\approx 10\%$ of the search space), or stop when reaching the global optimum.

The search trajectory is traced and filtered according to the basins of attraction: each solution belongs to one basin, so it is labelled by the corresponding local optimum. We then considered that a LS stays in the same basin, if the solutions s_t and s_{t+1} both belong to the same basin and the fitness increases: $f(s_t) \leq f(s_{t+1})$. Otherwise, the LS jumps from one basin to another one. In that manner, when accumulated from a number of independent runs of the LS, 10^4 in our experiments, it is possible to compare the empirical transition frequencies between basins with the corresponding edge weights of a given LON. Such a high number of independent runs is necessary because often times single trajectories go through few LOs and only once.

When a LS performs a transition between two basins that are not connected in the LON model, we add to the latter a virtual edge with weight equal to 0.0. Of course, we do not consider the LON edges that are not sampled, as it is not possible to compute the transition frequency between nodes that have not been visited. Finally, the LS under study are based on hill-climbing, thus the LON self-loops are also discarded.

4.2 Results

Figure 7 shows the scatter plots for the correlations between the edges weights of the different LON models and the empirical transition frequencies between basins.

Iterated Local Search. Our implementation is based on a steepest-ascent (best-improvement) hill-climbing, which, at each iteration, moves to the best neighboring so-

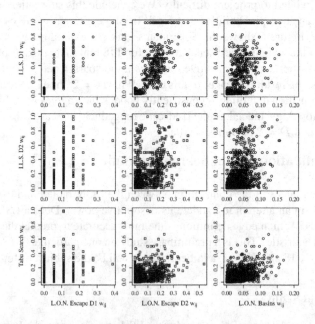

Fig. 7. Correlation (Spearman coefficient) between the edge weight and the empirical transition frequency of a local search. 10^4 independent runs on an NK-instance with $N = 18$, $K = 4$. From top to bottom, Iterated Local Search with 1- and 2-bit-flips, and Tabu Search have been tested. From left to right, the empirical frequencies are plotted against the corresponding edge weights according to the Escape $D \in \{1, 2\}$ and Basins definition, respectively. Only transitions between different basins are considered (no self-loops).

lution, and stops on a local optimum. The ILS perturbation is the k bit-flip mutation, which flips k bits at random, $k \in \{1, 2\}$ in the present study. A mutation is accepted as a new solution if its fitness value is strictly better than the current one. Intuitively, the ILS follows the same edges of the escape definition: from the LO, k bits are flipped; the the main difference lies in the acceptance criterion used by the ILS. Indeed (see Fig. 7), correlations are all significant. As expected, for the 1-bit-flip perturbation, Spearman coefficient is higher for escape edges with $D = 1$, and lower for $D = 2$-escape edges (0.49 for $D = 1$, 0.45 for $D = 2$, and 0.48 for basin edges). For the 2-bit-flips perturbation, the highest correlation is for the escape edges with $D = 2$, and the smallest for escape edges with $D = 1$ (0.41 for $D = 1$, 0.75 for $D = 2$, and 0.72 for basin edges). Notice how figures for basin edges and $D = 2$-escape edges are very similar.

Tabu Search. Our implementation uses the 1-bit-flip mutation, and, in order to obtain a better diversification, a set of moves are forbidden during the search; those tabu-moves are stored in a memory that lasts $\lambda = N = 18$ iterations. A tabu move is nonetheless accepted when it finds a new best-so-far solution. Referring to Figure 7, the correlation values are 0.27 for the $D = 1$-escape edges, 0.53 for $D = 2$, and 0.79 for the basin edges. All values are significant and positive, particularly so for the more interesting LON definitions: $D = 2$-escape edges and basin edges.

The result of this preliminary study is encouraging because it shows that the LON could capture the coarse-grained dynamics of a LS. In particular, the escape edge definition with $D = 2$ seems to be informative enough to correlate with the trajectories of a simple ILS, or a simple TS. Of course, further studies have to confirm this result for a larger class of LS, a broader number of instances, and on other problems, as to delineate the validity domain of the LON model.

5   Concluding Remarks

The local optima networks (LON) model is a mesoscopic representation of a problem search space, which deals with the local-optima basins of attractions as the meso-state level of description. In this contribution, an alternative definition of edges (*escape edges*), has been proposed. Our statistical analysis, on a set of NK-landscapes, shows that the escape edges are as informative as the original (basin- transition) edges. We reach this conclusion because, for both LON models, the analyzed network features (such as the clustering coefficient, disparity, correlation between in-strength and fitness of local optima, and path length to global optimum) are always consistent with the non-linearity of the problem (the K parameter), which tunes the landscape ruggedness. Indeed, the edges' weights are positively correlated between the different definitions.

We also present a preliminary analysis that aims at validating the model. We show that the dynamics of simple stochastic local search heuristics such as Iterated Local Search, or Tabu Search, tend to follow the edges of LONs according to the rate defined by the weights. The model validation in the present study is a first step. Starting from the work by Reidys and Stadler [9] on combinatorial fitness landscapes, we could conduct a spectral analysis of the LON model. From the adjacency matrix of the LON graphs, it should be possible to build a Markov Chain having the previously discussed transition probabilities. That would allow us to compare the stationary distribution of different

LON models of the same search space. In this case, an empirical assessment could be performed in a more informed way, because we could estimate the number of local search runs that are necessary to have a good sampling, as in a Monte Carlo method. The model validation could, then, go together with a prediction of the LS dynamics.

Overall, the present study opens up relevant perspectives. The escape edges definition will allow us to design a sampling methodology of the LONs. The enumeration of the basins of attraction is impossible on realistic search spaces. Therefore, the original definition of edges is restricted to study small search spaces. With the new definition, and through sampling of large networks such as breadth-first search, forest-fire or snowball sampling, we will be able to study the properties of LONs for real-world combinatorial search spaces. Our hope is that, by combining the LON model of the search dynamics and the ability to build LONs for large search spaces, we will be able to perform off-line parameter tuning of evolutionary algorithms and local search heuristics. Moreover, if the LON features are collected along the search process, on-line control of the parameters could also be achieved.

Acknowledgments. Fabio Daolio and Marco Tomassini gratefully acknowledge the Swiss National Science Foundation for financial support under grant number 200021-124578.

References

1. Barthélemy, M., Barrat, A., Pastor-Satorras, R., Vespignani, A.: Characterization and modeling of weighted networks. Physica A 346, 34–43 (2005)
2. Cahon, S., Melab, N., Talbi, E.G.: Paradiseo: A framework for the reusable design of parallel and distributed metaheuristics. Journal of Heuristics 10, 357–380 (2004)
3. Csardi, G., Nepusz, T.: The igraph software package for complex network research. Inter-Journal Complex Systems, 1695 (2006)
4. Glover, F., Laguna, M.: Tabu Search. Kluwer Academic Publishers, Norwell (1997)
5. Kauffman, S.A.: The Origins of Order. Oxford University Press, New York (1993)
6. Lourenço, H.R., Martin, O., Stützle, T.: Iterated local search. In: Handbook of Metaheuristics. International Series in Operations Research & Management Science, vol. 57, pp. 321–353. Kluwer Academic Publishers (2002)
7. Newman, M.E.J.: The structure and function of complex networks. SIAM Review 45, 167–256 (2003)
8. Ochoa, G., Verel, S., Tomassini, M.: First-Improvement vs. Best-Improvement Local Optima Networks of NK Landscapes. In: Schaefer, R., Cotta, C., Kołodziej, J., Rudolph, G. (eds.) PPSN XI. LNCS, vol. 6238, pp. 104–113. Springer, Heidelberg (2010)
9. Reidys, C., Stadler, P.: Combinatorial landscapes. SIAM Review 44(1), 3–54 (2002)
10. Tomassini, M., Verel, S., Ochoa, G.: Complex-network analysis of combinatorial spaces: The NK landscape case. Phys. Rev. E 78(6), 066114 (2008)
11. Verel, S., Ochoa, G., Tomassini, M.: Local optima networks of NK landscapes with neutrality. IEEE Transactions on Evolutionary Computation 15(6), 783–797 (2011)

Visual Analysis of Population Scatterplots

Evelyne Lutton[1], Julie Foucquier[2], Nathalie Perrot[2],
Jean Louchet[3], and Jean-Daniel Fekete[1]

[1] AVIZ Team, INRIA Saclay - Ile-de-France,
Bat 490, Université Paris-Sud, 91405 ORSAY Cedex, France
{Evelyne.Lutton,Jean-Daniel.Fekete}@inria.fr
[2] UMR782 Génie et Microbiologie des Procédés Alimentaires,
AgroParisTech, INRA, 78850 Thiverval-Grignon, France
{Julie.Foucquier,nathalie.perrot}@grignon.inra.fr
[3] Artenia, 24 rue Gay-Lussac, 92320 Châtillon, France
Jean.Louchet@gmail.com

Abstract. We investigate how visual analytic tools can deal with the huge amount of data produced during the run of an evolutionary algorithm. We show, on toy examples and on two real life problems, how a multidimensional data visualisation tool like ScatterDice/GraphDice can be easily used for analysing raw output data produced along the run of an evolutionary algorithm. Visual interpretation of population data is not used very often by the EA community for experimental analysis. We show here that this approach may yield additional high level information that is hardly accessible through conventional computation.

Keywords: Optimisation, Artificial Evolution, Genetic algorithms, Visual Analytics, Experimental analysis of algorithms, fitness landscape visualisation.

1 Introduction

Experimental analysis is an important aspect of research in Evolutionary Computation, both in the development of new algorithms and in the understanding of the internal mechanisms of artificial evolution. It is particularly difficult to deal with the huge amount of information that can be collected during the run of an evolutionary algorithm, as the information is usually highly multidimensional, and combines continuous, discrete and symbolic data.

Visualisation utilities associated with available EA software usually display fitness curves, convergence diagrams, and various statistics, but they more rarely allow to visualise raw genomic data. The importance of efficient visualisation tools for EAs is not a new issue [18], and various solutions have been proposed. The methods are commonly split into two types of visualisation tools: on-line tools, that display a set of monitoring curves during an EA run (e.g. fitness of the best individual, statistics, diversity), and off-line tools, that do a "post-mortem" analysis with more sophisticated results. However, important, difficult issues remain unsolved:

- How to visualise an individual. This task is particularly complex when the search space is large and multidimensional, or when the genome combines symbolic and

J.-K. Hao et al. (Eds.): EA 2011, LNCS 7401, pp. 61–72, 2012.

numerical (discrete and/or continuous) values. The particular case of genetic programming has also been considered[7]. Existing solutions are usually problem dependent, and may call for the display of phenotypes (signals, images, sounds, graphs, networks, etc ...).

– How to visualise a population, i.e. a possibly large set of multidimensional points. The ability to efficiently visualise fitness landscapes is still a challenge: approaches like fitness-distance correlation only give a partial answer.
– How to visualise the history and evolutionary mechanisms. This issue is important, as visualising various statistics about an evolving population may not be enough to understand some of the complex or hidden mechanisms of EAs, like the action of operators. Being able to follow the transmission of genetic material inside a population has been partially addressed by schemata analysis, however there is still a strong need when dealing with continuous landscapes.
– How to visualise the result in non-standard EAs. A good example of this is the case of multi-objective EAs, as the growing size of the problem EA are able to solve lead to outputs made of large, high dimensional Pareto datasets. There is a strong need of efficient visualisation tools, that may help to monitor (on line or even interactively) multidimensional Pareto fronts [13].

In this paper, we investigate recent tools developed by the visual analytics community, that may provide efficient and generic answers to some of the previous challenges. The paper is organised as follows. A short review of existing visualisation systems for EAs is given in Section 2. Section 3 presents the ScatterDice / GraphDice tool. An analysis of population clouds using ScatterDice/GraphDice is developed in Section 4 for some classical test functions. Tests are based on the EASEA language[4][1]. Two real life examples are then presented in Section 5, and as a conclusion Section 6 sketches future developments for a GraphDice version adapted to EA visualisation.

2   Visualisation of EA Data

2.1   On-line Visualisation

Almost any evolutionary software proposes nowadays its own on-line display facilities. It is often reduced to visualise how the best fitness value evolves along generations. We give below some examples that provide additional features.

Bedau and Bullock[1,3] show the importance of tracking evolutionary activity via plots that visualise the history at different levels. For instance, genotype's activity corresponds to the frequence of a given genotype in a population, which appears, increases or decreases along generations, forming what they call "waves".

Pohlheim[15] proposed in 1999 a visualisation system adapted to the Genetic and Evolutionary Algorithm Toolbox for Matlab - GEATbx[14]. His system allows various visualisation modes, and gives for instance the current state of a population (one generation), visualises a run (all the generations), or different runs for comparisons. Additionally he copes with the problem of visualising high-dimensional data using

---

[1] The software is available at http://sourceforge.net/projects/easea/

multidimensional scaling (reduction to a 2D visualisation that preserves distance relationships), and uses a 2D representation to show the paths followed by the best individual of every generation.

Kerren and Egger in 2005 [6] developed a Java-based on-line visualisation tool, EAVis, in which it is possible to embed an application-dependent Phenotype View.

Collins in 2003 [5] provided a survey chapter on this topic and identified some directions for future research. His main conclusion concerns the strong need for flexible visualisation environments, as he considered that current solutions were still too problem dependent.

2.2   Off-Line Visualisation

Off-line visualisation systems allow displaying more data, including multidimensional data, which is one of the important issues in current visualisation systems. Spears [19] provided in 1999 a brief overview of multidimensional visualisation techniques for visualising EAs, such as the use of colour, glyphs, parallel coordinates or coordinates projections. For discrete data, Routen in 1994 [16] suggested to adopt a modified Hinton diagram[2] in which a rectangle represents a chromosome and its size is determined by the fitness of the chromosome. Let us give below a list of some off-line systems:

- William Shine and Christoph Eick[18] describe the features of a GA-visualization environment that uses quadcodes to generate search space coverage maps, employs 2D-distance maps to visualize convergence, and uses contour maps to visualize fitness.
- The VIS system [21] proposed in 1999 allows a navigation at various levels of detail, and manages transitions between related data levels. The visualisation of the most detailed level is based on ad-hoc representations (bar codes, colors, alleles frequencies) but does not allow visualising multidimensional continuous genomes.
- Emma Hart [10] proposed GAVEL in 2001, an off-line visualisation system adapted to generational GAs, that provides a means to understand how crossover and mutation operations assemble the optimal solution, and a way to trace the history of user-selected sets of alleles. It allows a display of the complete history across all generations of chromosomes, individual genes, and schemata.
- Marian Mach [12] presented in 2002 a simple and interactive "post-mortem" GA visualising tool focused on the visualisation of multidimensional data via 1D projections.

Annie Wu [21], who developed the VIS system, gave the following list of desirable tasks for visualisation systems: *(a)* to examine individuals and their encodings in detail, *(b)* to trace the source and survival of building blocks or partial solutions, *(c)* to trace family trees to examine the effects of genetic operators, *(d)* to examine populations for convergence, speciation, etc, *(e)* to trace gross population statistics and trends to move freely in time and through populations.

---

[2] A Hinton diagram provides a qualitative display of the values in a matrix. Each value is represented by a square whose size is related to the magnitude, and color indicates sign.

The genome representation issue, item *(a)*, is perhaps the most complex one, and as we have seen above, various solutions have been proposed, depending if we are dealing with continuous or discrete genomes. For Genetic Programming, the question is even more complex: a solution proposed by Jason Daida [7] in 2005 consists in visualizing big tree structures as large sized graphs.

The issue addressed by the GAVEL system, that appears as very challenging in off-line systems and that is mentioned by Annie Wu as items *(c)* and *(d)*, is to be able to trace the history of individuals. Spears [19] mentioned also that being able to track the origin of the fittest individual per generation is a very important issue for parameter tuning.

The question of family trees visualisation has more recently been considered by Zbigniew Walczak in 2005 in a short chapter [20] where he proposed to visualise evolutionary processes using graph drawing software.

3   ScatterDice / GraphDice

Visual analytics is a multidisciplinary field that integrates various sophisticated computational tools with innovative interactive techniques and visual representations to facilitate human interpretation of raw data. We present in this section a tool developed by researchers in this field, that seems to answer in a generic way to some needs identified in the previous sections.

ScatterDice[8] is a multidimensional visual exploration tool, that enables the user to navigate in a multidimensional set via simple 2D projections, organised as scatterplot matrices. The visual coherence between various projections is based on animated 3D transitions. A scatterplot matrix presents an overview of the possible configurations, thumbnails of the scatterplots, and support for interactive navigation in the multidimensional space. Various queries can be built using bounding volumes in the dataset, sculpting the query from different viewpoints to become more and more refined. Furthermore, the dimensions in the navigation space can be reordered, manually or automatically, to highlight salient correlations and differences among them[3].

A recent evolution of ScatterDice using the same principles but with many additional features is GraphDice[2]. It allows reading the same type of data (.csv files), and other more sophisticated formats, as it also embeds graph visualisation utilities[4].

4   Analysing Successive Populations

ScatterDice or GraphDice can be used to visualise data collected during the run of an EA[11]. At each generation, the content of the current population can be written into a ".csv" file as on figure 1, creating what can be called a "cloud" of successive populations. This cloud of multidimensional points is visualised using ScatterDice or GraphDice, to produce various, sometimes unusual, viewpoints.

---

[3] A demo of ScatterDice can be launched from `http://www.aviz.fr/~fekete/scatterdice/`, it accepts standard .csv files (although it may be necessary to add a second line after the header giving the data type for each column - INT, STR, REAL, etc).

[4] A demo of GraphDice is also accessible at `http://www.aviz.fr/graphdice/`

```
Generation;Fitness;x[0];sigma[0];x[1];sigma[1]
INT;DOUBLE;DOUBLE;DOUBLE;DOUBLE;DOUBLE
0;3,75447;-0.12508;0.195626;0.524069;0.255402
0;1,17484;-0.573358;0.142053;0.924887;0.392851
0;2,28066;-0.533583;0.183641;0.546523;0.461643
0;1,92319;-0.70052;0.338246;0.582533;0.406443
0;2,75539;0.784538;0.182157;-0.940648;0.383136
0;3,08894;-0.770051;0.190012;-0.840338;0.359992
0;2,30766;0.380979;0.124581;0.0379446;0.469388
0;3,30957;-0.704403;0.453222;0.208484;0.182612
...
```

Fig. 1. A simple .csv file collected during a run (minimisation of the 2D Weierstrass function $H = 0.2$)

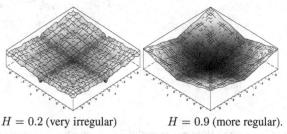

$H = 0.2$ (very irregular) $H = 0.9$ (more regular).

Fig. 2. 2D Weierstrass functions with Hölder Exponent H

The test functions used in the experiments below are the following:

– Weierstrass functions (see figure 2) defined in a space of dimension 2, of Hölder exponents $H = 0.2$ (very irregular) and $H = 0.9$ (more regular).

$$f(x,y) = \sum_{n=-\infty}^{+\infty} 2^{-nH}(1 - \cos 2^n x) + \sum_{n=-\infty}^{+\infty} 2^{-nH}(1 - \cos 2^n y)$$

– Rosenbrock function[5] in a space of dimension 10.

$$f(x_1, ..., x_{10}) = \sum_{n=1}^{9} 100(x_i^2 - x_{i+1})^2 + (1 - x_i)^2$$

The genetic engine is a simple generational algorithm on R^n using tournament selection, geometric (barycentric) crossover and log-normal self-adaptive mutation. Additional dimensions are thus considered in the search space, the σ_i values, that represent the mutation radius for each coordinate x_i. The population size is 100 and the algorithm runs for 100 generations. This genetic engine is available in the sample programs (weierstrass.ez) distributed with the EASEA software.

A visualisation of the population cloud for the 2D Weierstrass function with dimension 0.2 is given on figure 3. The scatterplot matrix on the left of the figure, gives an overview of the possible visualisations. The columns and lines of this matrix can be dragged and dropped as wished by the user. A default order is proposed, based on an

---

[5] See for instance http://en.wikipedia.org/wiki/Rosenbrock_function

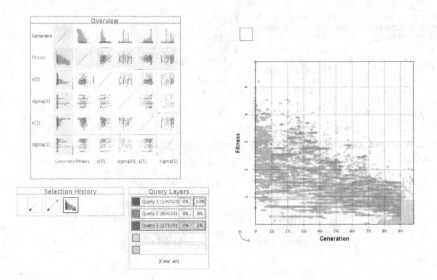

Fig. 3. 2D Weierstrass function of Hölder exponent 0.2. Scatterplot and Fitness versus generation view. Red points correspond to the first 10 generations, yellow points, to the 10 last ones, and green points, to the best fitness areas.

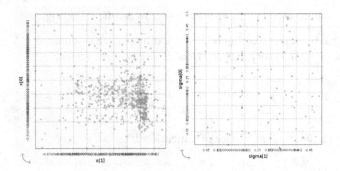

Fig. 4. 2D Weierstrass function $H = 0.2$. Left, projection on the 2D plane (x_0, x_1). Right, parameters σ_0 and σ_1. Points in red belong to generations 0 to 10, points in yellow to 90 to 100.

algorithm that reduces "clutter"[2]. On the right, a detailed view is given, corresponding to the selected cell of the matrix (hightlighted in red), on which some queries have been visualised: in red, the points corresponding to the first 10 generations of the run, in yellow, the last 10 generations, and in green, the best fitness points. The queries are organised as layers, the "Query layers" window gives the details of the three queries, with some additional measurements (percentage of selected points, and percentage of selected edges if a graph is visualised : GraphDice actually considers a set of points as a degenerate graph, made of a collection of nodes with no edges). Bottom left, a "Selection History" window shows how the queries have been sculpted: queries 1 and 2 have been activated on the Generation versus Generation view, i.e. the top left plot

of the scatterplot matrix, while query 3 has been made on the fitness versus generation plot. On the extreme right of the window, a toolbar proposes various visualisation
options (the "Show" window), for instance "Grid" activates a grid on the dataset, "Labels" allow to display attributes attached to a point. It is thus possible for instance to
identify different runs of an EA using a label, and use this option to separate the data
when needed. "Hull" displays a convex hull for each query, and "Zoom" activates an
automatic zoom focused on the selected data.

The Fitness versus Generation plot, that displays the whole set of individuals generated along the evolution, provides additional information about the distribution of successive populations. When observed from a different viewpoint, for instance according
to x_0 and x_1 or to σ_0 and σ_1 like in figure 4, it can be noticed that the population diversity decreases slowly, and converge toward a point of the 2D plane, while stabilising on
some mutation parameters. The first generations (red points of the query 1) are spread
in a rather uniform way on the whole search space, while the last ones (yellow points)
are concentrated in the areas of best fitness (green points). A green point appears rather
early (see generation 12 on the Fitness versus Generation plot), but green points start to
multiplicate rather late (from generation 68) in the evolution.

Figure 5 gives the same view as figure 3 but for a 2D Weierstrass function of dimension 0.9. This function is much more regular than the previous one. Visually, it seems
obvious that the population is able to converge more rapidly.

Figure 6 gives an overview of the visualisation window for a dimension 10 space.
Once again, as the function is more regular, the population seems to converge rapidly,
even if it uses the same parameters as for the Weierstrass functions, in a search space
with higher dimension.

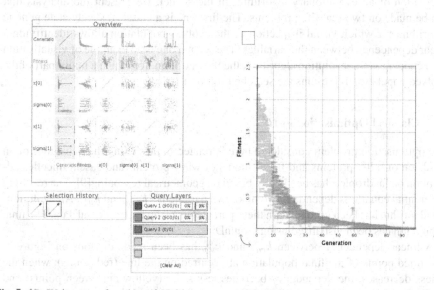

Fig. 5. 2D Weierstrass function of Hölder exponent 0.9. Yellow points correspond to the first
generations, red points to the last generations.

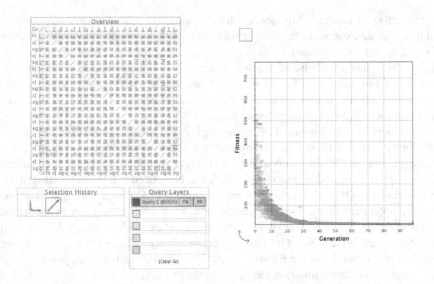

Fig. 6. Rosenbrock function in dimension 10. Red points correspond to the 10 last generations.

5   Analysing an Evolved Population

GraphDice can also be used a more conventional way, in order to visualise the final population of an evolutionary algorithm. In the sequel, we present the analysis that can be made on two real life problems. The first one is a classical steady-state genetic algorithm, for which visual inspection of the evolved population allows identifying a linear dependency between the variables. The second one is an example of visualisation for a cooperative-coevolution algorithm, the fly algorithm, for which a major part of the evolved population represents the searched solution.

5.1   Classical Optimisation in R^2

The optimisation problem considered here is related to modelling the behaviour of an emulsion of milk, proteins and fat. A model of whey protein[6] and casein micelles adsorbtion on fat droplets has been built, based on geometrical reasoning[9]. The resulting differential equations depend on two unknown parameters k_{cas} and k_{wp}. An evolutionary algorithm has been used to learn these parameters using experimental data. The final population has been visualised using GraphDice.

A linear dependency between k_{cas} and k_{wp} has been made obvious on Figure 7: best fitted points of the final population are centred on the line (red points); when the fitness decreases, the dependency becomes less strict (yellow and green points) and the corresponding points are distributed around the diagonal line. It seems thus more convenient to deal with a one dimensional problem, i.e. to find the best ratio k_{cas}/k_{wp}.

---

[6] Whey is left over when milk coagulates and contains everything that is soluble from milk.

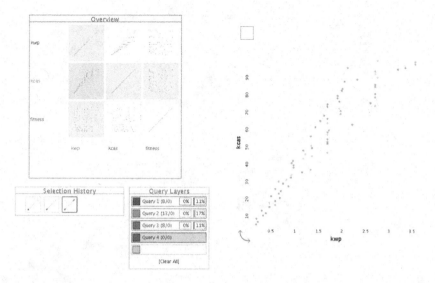

Fig. 7. Red points correspond to the 11% best fitness individuals, yellow points, to the 17% "next" best points, and green points, to the 11% worst fitness points. Queries have been interactively sculpted on the fitness versus fitness view. On the *kcas* versus *kwp* view, a linear dependence between the two parameters has been made obvious.

This experimental evidence sheds a new light on the *a priori* geometrical model, which needs now to be refined.

5.2 Cooperative Coevolution of a Set of 3D Points

In the computer vision domain, Cooperative Coevolution algorithms (CCEAs) have been applied to stereovision, to produce the fly algorithm[17]. This algorithm evolves a population of 3-D points, the flies, so that the population matches the shapes of the objects on the scene. It is a cooperative coevolution in the sense that the searched solution is represented by the whole population rather than by the single best individual.

An individual of the population, a "fly", is defined as a 3-D point with coordinates (x,y,z). If the fly is on the surface of an opaque object, then the corresponding pixels in the two images will normally have highly similar neighbourhoods. Conversely, if the fly is not on the surface of an object, their close neighbourhoods will usually be

Fig. 8. Visualisation of results by reprojection on the stereo pair: flies are in pink

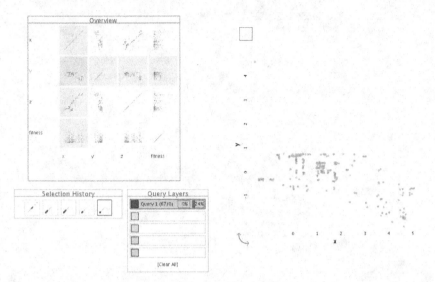

Fig. 9. Front view of the flies cloud: furniture structures are visible. Red points are the best fitted flies.

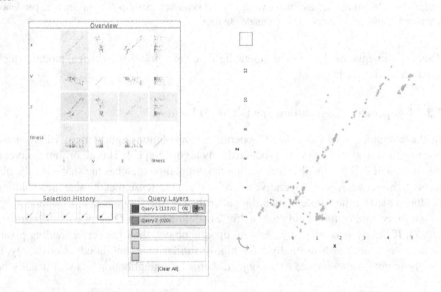

Fig. 10. View from above: the direction of the wall is visible

poorly correlated. The fitness function exploits this property and evaluates the degree of similarity of the pixel neighbourhoods of the projections of the fly, giving higher fitness values to those probably lying on objects surfaces.

GraphDice allows visualising a population of flies, and rapidly provides various viewpoints on the evolved data set (figures 9 and 10), in addition to *ad hoc* visualisations (figure 8).

6   Conclusion and Future Work

We have shown above how GraphDice allows fast visual inspection of raw EA data. For instance it is easy to visualise the algorithm's exploration capability and population diversity, and to localise possible visually obvious dependencies between variables.

Some features of GraphDice are particularly useful to make it a good candidate for a generic answer to the issues identified in section 1: The visualisation scheme is simple and easy to use (2D scatterplots organised in a matrix). Interactions and graphic queries allow a fast navigation in the datased. Data exchange is performed via simple csv files.

Our intention is now to adapt this general purpose visualisation tool to the specific needs of EA analysis. The following issues will guide future developments of a GraphDice version adapted to EAs:

- Tests have been performed on relatively small data sets (up to 100x100 individuals in 10 dimensional space). The scalability issue will be tested more extensively.
- Various usual statistics (per generation, per fitness level) and query-based statistics will be implemented, including comparison of distributions (p-values).
- On-line visualisation issues will also be considered.

References

1. Bedau, M.A., Joshi, S., Lillie, B.: Visualizing waves of evolutionary activity of alleles. In: Proceedings of the 1999 GECCO Workshop on Evolutionary Computation Visualization, pp. 96–98 (1999)
2. Bezerianos, A., Chevalier, F., Dragicevic, P., Elmqvist, N., Fekete, J.-D.: Graphdice: A system for exploring multivariate social networks. Computer Graphics Forum (Proc. EuroVis 2010) 29(3), 863–872 (2010)
3. Bullock, S., Bedau, M.A.: Exploring the dynamics of adaptation with evolutionary activity plots. Artif. Life 12, 193–197 (2006)
4. Collet, P., Lutton, E., Schoenauer, M., Louchet, J.: Take it EASEA. In: Schoenauer, M., Deb, K., Rudolph, G., Yao, X., Lutton, E., Merelo, J.J., Schwefel, H.-P. (eds.) PPSN 2000. LNCS, vol. 1917, pp. 891–901. Springer, Heidelberg (2000)
5. Collins, T.D.: Visualizing evolutionary computation, pp. 95–116. Springer-Verlag New York, Inc., New York (2003)
6. Kerren, A.: Eavis: A visualization tool for evolutionary algorithms. In: Proceedings of the IEEE Symposium on Visual Languages and Human-Centric Computing (VL/HCC 2005), pp. 299–301 (2005)
7. Daida, J., Hilss, A., Ward, D., Long, S.: Visualizing tree structures in genetic programming. Genetic Programming and Evolvable Machines 6, 79–110 (2005)
8. Elmqvist, N., Dragicevic, P., Fekete, J.-D.: Rolling the dice: Multidimensional visual exploration using scatterplot matrix navigation. IEEE Transactions on Visualization and Computer Graphics (Proc. InfoVis 2008) 14(6), 1141–1148 (2008)
9. Foucquier, J., Gaucel, S., Surel, C., Anton, M., Garnier, C., Riaublanc, A., Baudrit, C., Perrot, N.: Modelling the formation of the fat droplets interface during homogenisation in order to describe texture. In: ICEF, 11th International Congress on Engineering and Food, Athens, Greece, May 22-26 (2011), http://www.icef11.org/
10. Hart, E., Ross, P.: Gavel - a new tool for genetic algorithm visualization. IEEE Trans. Evolutionary Computation 5(4), 335–348 (2001)

11. Lutton, E., Fekete, J.-D.: Visual analytics of ea data. In: Genetic and Evolutionary Computation Conference, GECCO 2011, Dublin, Ireland, July 12-16 (2011)
12. Mach, Z., Zetakova, M.: Visualising genetic algorithms: A way through the Labyrinth of search space. In: Sincak, P., Vascak, J., Kvasnicka, V., Pospichal, J. (eds.) Intelligent Technologies - Theory and Applications, pp. 279–285. IOS Press, Amsterdam (2002)
13. Parmee, I.C., Abraham, J.A.R.: Supporting implicit learning via the visualisation of coga multi-objective data. In: CEC 2004, Congress on Evolutionary Computation, June 19-23, vol. 1, pp. 395–402 (2004)
14. Pohlheim, H.: Geatbx - genetic and evolutionary algorithm toolbox for matlab, http://www.geatbx.com/
15. Pohlheim, H.: Visualization of evolutionary algorithms - set of standard techniques and multidimensional visualization. In: GECCO 1999 - Proceedings of the Genetic and Evolutionary Computation Conference, San Francisco, CA, pp. 533–540 (1999)
16. Routen, T.W.: Techniques for the visualisation of genetic algorithms. In: The First IEEE Conference on Evolutionary Computation, vol. II, pp. 846–851 (1994)
17. Sapin, E., Louchet, J., Lutton, E.: The fly algorithm revisited: Adaptation to cmos image sensor. In: ICEC 2009, International Conference on Evolutionary Computation, Madeira, Portugal, October 5-7 (2009)
18. Shine, W., Eick, C.: Visualizing the evolution of genetic algorithm search processes. In: Proceedings of 1997 IEEE International Conference on Evolutionary Computation, pp. 367–372. IEEE Press (1997)
19. Spears, W.M.: An overview of multidimensional visualization techniques. In: Collins, T.D. (ed.) Evolutionary Computation Visualization Workshop, Orlando, Florida, USA (1999)
20. Walczak, Z.: Graph-Based Analysis of Evolutionary Algorithm. In: Klopotek, M., Wierzchon, S., Trojanowski, K. (eds.) Intelligent Information Processing and Web Mining. AISC, vol. 31, pp. 329–338. Springer, Heidelberg (2005)
21. Wu, A.S., De Jong, K.A., Burke, D.S., Grefenstette, J.J., Ramsey, C.L.: Visual analysis of evolutionary algorithms. In: Proceedings of the 1999 Conference on Evolutionary Computation (CEC 1999), pp. 1419–1425. IEEE Press (1999)

An On-Line On-Board Distributed Algorithm
for Evolutionary Robotics

Robert-Jan Huijsman, Evert Haasdijk, and A.E. Eiben

Dept. of Computer Science, Vrije Universiteit Amsterdam, The Netherlands
rjhuijsman@gmail.com, {e.haasdijk,a.e.eiben}@vu.nl
http://www.cs.vu.nl/ci/

Abstract. Imagine autonomous, self-sufficient robot collectives that can adapt their controllers autonomously and self-sufficiently to learn to cope with situations unforeseen by their designers. As one step towards the realisation of this vision, we investigate on-board evolutionary algorithms that allow robot controllers to adapt without any outside supervision and while the robots perform their proper tasks. We propose an EVAG-based on-board evolutionary algorithm, where controllers are exchanged among robots that evolve simultaneously. We compare it with the $(\mu + 1)$ ON-LINE algorithm, which implements evolutionary adaptation inside a single robot. We perform simulation experiments to investigate algorithm performance and use parameter tuning to evaluate the algorithms at their *best possible* parameter settings. We find that distributed on-line on-board evolutionary algorithms that share genomes among robots such as our EVAG implementation effectively harness the pooled learning capabilities, with an increasing benefit over encapsulated approaches as the number of participating robots grows.

Keywords: evolutionary robotics, on-line evolution, distributed evolution.

1 Introduction

The work presented in this paper is inspired by a vision of autonomous, self-sufficient robots and robot collectives that can cope with situations unforeseen by their designers. An essential capability of such robots is the ability to adapt –evolve, in our case– their controllers in the face of challenges they encounter in a hands-free manner, "the ability to learn control without human supervision," as Nelson *et al.* put it [13]. In a scenario where the designers cannot predict the operational circumstances of the robots (e.g, an unknown environment or one with complex dynamics), the robots need to be deployed with roughly optimised controllers and the ability to evolve their controllers autonomously, on-line and on-board.

This contrasts with the majority of evolutionary robotics research, which focusses on *off-line* evolution, where robot controllers are developed -evolved- in a separate training stage before they are deployed to tackle their tasks in earnest.

When dealing with multiple autonomous robots, one can distinguish three options to implement on-line evolution [7]:

Encapsulated. Each robot carries an isolated and self-sufficient evolutionary algorithm, maintaining a population of genotypes inside itself;

J.-K. Hao et al. (Eds.): EA 2011, LNCS 7401, pp. 73–84, 2012.

Distributed. Each robot carries a single genome and the evolutionary process takes place by exchanging genetic material between robots;

Hybrid. Which combines the above two approaches: each robot carries multiple genomes and shares these with its peers.

In this paper, we compare instances of each of these three schemes: the encapsulated $(\mu + 1)$ ON-LINE algorithm, the distributed Evolutionary Agents algorithm (EVAG) [9] and a hybrid extension of EVAG.

One of EVAG's distinguishing features is that it employs the newscast algorithm [8] to exchange genomes between peers (in our case: robots) and to maintain an overlay network for peer-to-peer (robot-to-robot) communication. Newscast-based EVAG has proved very effective for networks of hundreds or even thousands of peers, but in the case of swarm robotics, network sizes are likely to be much smaller. Therefore, it makes sense to compare the efficacy of newscast-based EVAG with a panmictic variant where the overlay network is fully connected.

It is well known that the performance of evolutionary algorithms to a large extent depends on their parameter values [11]. To evaluate the algorithms at their *best possible* parameter setting, we tune the algorithm parameters with REVAC, an evolutionary tuning algorithm specifically designed for use with evolutionary algorithms [10].

Summarising, the main question we address in this paper is: how do the three algorithms compare in terms of performance and can we identify circumstances in which to prefer one of the three schemes over the others? Secondly, we investigate how EVAG's newscast population structure influences its performance compared to a panmictic population structure. Thirdly, we briefly consider the sensitivity of the algorithms to parameter settings.

2   Related Work

The concept of on-board, on-line algorithms for evolutionary robotics was discussed as early as 1995 in [15], with later research focussed on the 'life-long learning' properties of such a system [20].

A distributed approach to on-line, on-board algorithms was first investigated as 'embodied evolution' in [21], where robots exchange single genes at a rate proportional to their fitness with other robots that evolve in parallel. Other work on on-line evolution of robot controllers is presented in [4] that describes the evolution of controllers for activating hard-coded behaviours for feeding and mating. In [1], Bianco and Nolfi experiment with open-ended evolution for robot swarms with self-assembling capabilities and report results indicating successful evolution of survival methods and the emergence of multi-robot individuals with co-ordinated movement and co-adapted body shapes.

The $(\mu + 1)$ ON-LINE algorithm – this paper's exemplar for the encapsulated approach – has been extensively described in [7] and [2], where it was shown to be capable of evolving controllers for a number of tasks such as obstacle avoidance, phototaxis and patrolling. [5] uses encapsulated evolution to evolve spiking circuits for a fast forward task. Encapsulated on-line evolution as a means for continuous adaptation by using genetic programming is suggested in [16].

The distributed approach to on-line evolutionary robotics has a clear analogy with the field of parallel evolutionary algorithms, in particular to the fine-grained approach, where each individual in the population has a processor of its own. The primary distinguishing factor among fine-grained parallel algorithms is their population structure, with small-world graphs proving competitive with panmictic layouts [6].

The hybrid scheme can be implemented as what in parallel evolutionary algorithms is known as the island model: the population is split into several separately evolving sub-populations (the islands), that occasionally exchange genomes. This approach is used in [19] and [4]. A variant where the robots share a common hall of fame is implemented in [12]. As will become apparent, we take a slightly different approach where genome exchange between sub-populations is the norm.

3 Algorithms

Autonomous on-line adaptation poses a number of requirements that regular evolutionary algorithms don't necessarily have to contend with. We take a closer look at two especially relevant considerations.

To begin with, fitness must be evaluated *in vivo*, i.e., the quality of any given controller must be determined by actually using that controller in a robot as it goes about its tasks. Such real-life, real-time fitness evaluations are inevitably very noisy because the initial conditions for the genomes under evaluation vary considerably. Whatever the details of the evolutionary mechanism, different controllers will be evaluated under different circumstances; for instance, the nth controller will start at the final location of the $(n - 1)$th one. This leads to very dissimilar evaluation conditions and ultimately to very noisy fitness evaluations. To address this issue, the algorithms we investigate here implement re-evaluation: whenever a new evaluation period commences, the robot can choose (with a probability ρ) not to generate a new individual but instead re-evaluate an existing individual to refine the fitness assessment and so combat noise.

The second issue specific to on-line evolution is that, in contrast to typical applications of evolutionary algorithms, the best performing individual is not the most important factor when applying on-line adaptation. Remember that controllers evolve as the robots go about their tasks; if a robot continually evaluates poor controllers, that robot's *actual* performance will be inadequate, no matter how good the best known individuals as archived in the population. Therefore, the evolutionary algorithm must converge rapidly to a good solution (even if it is not the best) and search prudently: it must display a more or less stable but improving level of performance throughout the continuing search.

3.1 $(\mu + 1)$ ON-LINE

The $(\mu + 1)$ ON-LINE algorithm is based on the classical $(\mu + 1)$ evolutionary strategy [17] with modifications to handle noisy fitness evaluations and promote rapid convergence. It maintains a population of μ individuals within each robot and these are evaluated in a time-sharing scheme, using an individual's phenotype as the robot's controller for a specified number of time units. A much more detailed description of $(\mu+1)$ ON-LINE is given in [7].

3.2 EVAG

EVAG was originally presented in [9] as a peer-to-peer evolutionary algorithm for parallel tackling of computationally expensive problems, with the ability to harness a large number of processors effectively. The analogies between parallel evolutionary algorithms and a swarm of robots adapting to their environment and tasks in parallel make EVAG a suitable candidate for an on-board, on-line distributed evolutionary algorithm for evolutionary robotics.

The basic structure of EVAG is straightforward and similar to a $1 + 1$ evolution strategy: each peer (robot) maintains a record of the best solution evaluated by that peer up until that point – the champion. For every new evaluation a new candidate is generated, using crossover and mutation; if the candidate outperforms the current champion it replaces the champion.

The basic definition of EVAG leaves many decisions open to the implementer, such as the choice of recombination and mutation operators and the details of parent selection (other than that it should select from peers' champions). Because we are interested in the effects of the distributed nature of the algorithm rather than those due to, say, different recombination schemes, we have chosen our evolutionary operators to match the $(\mu + 1)$ ON-LINE algorithm. As a result, the only difference between EVAG and $(\mu + 1)$ ON-LINE (with $\mu = 1$) lies in the exchange of genomes between robots and using a cache of received genomes rather than only a locally maintained population when selecting parents.

In this light, the extension of regular EVAG to a hybrid form is a straightforward one: rather than maintaining only a single champion on-board, the robots now maintain a population of μ individuals locally. With $\mu = 1$, this implements the distributed scheme, with $\mu > 1$, it becomes a hybrid implementation. With the cache of received genomes disabled, it boils down to $(\mu + 1)$ ON-LINE, our encapsulated algorithm. The pseudo code in algorithm 1 illustrates the overlap between these three implementations.

EVAG normally uses newscast [8] to exchange solutions efficiently while maintaining a low number of links between peers: each robot locally maintains a cache of recently received genomes. Periodically, each robot randomly takes one of the genomes in its cache and contacts the robot at which that genome originated. These two robots then exchange the contents of their cache. When needed, parents are selected from the union of this cache and the local champion using binary tournament selection. Because [8] showed that with this update scheme, picking a genome randomly from the cache of received genomes is all but equivalent to picking one randomly from the entire population, this assures that the binary tournament takes place as if the contestants were randomly drawn from the combined population across all robots.

Earlier research showed very promising results for EVAG [9], but these results were obtained using thousands of nodes, while evolutionary robotics generally takes place with group sizes of no more than a few dozen robots. To investigate if EVAG's newscast overlay network remains efficient in these smaller populations we evaluate not only the standard newscast-based EVAG, but also an EVAG variant that uses a panmictic population structure where a robot's choice of genomes for the binary tournament parent selection is truly uniform random from the entire population across all robots. This variant of EVAG requires full connectivity among peers (robots) and is therefore not

```
for i ← 1 to μ do                                          // Initialisation
    population[i] ← CreateRandomGenome();
    population[i].σ ← σ_initial; // Mutation step size, updated cf. [2]
    population[i].Fitness ← RunAndEvaluate(population[i]);
end
for ever do                                         // Continuous adaptation
    if random() < ρ then // Don't create offspring, but re-evaluate
    selected individual
        Evaluatee ← BinaryTournament(population);
        Evaluatee.Fitness ← (Evaluatee.Fitness + RunAndEvaluate(Evaluatee)) / 2;
        // Combine re-evaluation results through exponential
        moving average
    else       // Create offspring and evaluate that as challenger
        ParentA ← BinaryTournament(pool of possible parents);
        ParentB ← BinaryTournament(pool of possible parents - parentA);
        if random() < crossover Rate then
            Challenger ← AveragingCrossover(ParentA, ParentB);
        else
            Challenger ← ParentA;
        end
        if random() < mutation Rate then
            Mutate(Challenger);            // Gaussian mutation from N(0, σ)
        end
        Challenger.Fitness ← RunAndEvaluate(Challenger);
        if Challenger.Fitness > population[μ].Fitness then       // Replace last
        (i.e. worst) individual in population w. elitism
            population[μ] ← Challenger;
            population[μ].Fitness ← Challenger.Fitness;
        end
    end
    Sort(population);
end
```

Algorithm 1. The on-line evolutionary algorithm. For $(\mu + 1)$ ON-LINE, the pool of possible parents is the on-board population of size μ; for both regular and hybrid EVAG, it is the union of individuals received from the robot's peers through newscast and the on-board population (with $\mu = 1$ for standard EVAG).

suitable for use in truly large-scale applications. However, evaluation of the panmictic structure compared to that of newscast is interesting, since it allows us to determine the performance penalty of using of a peer-to-peer approach.

4   Experiments

To investigate the performance of the algorithms and their variants we conduct experiments with simulated e-pucks in the ROBOROBO[1] environment. The experiments have

---

[1] http://www.lri.fr/~bredeche/roborobo/

the robots running their own autonomous instance of $(\mu + 1)$ ON-LINE, EVAG or its hybrid extension EVAG, governing the weights of a straightforward perceptron neural net controller with hyperbolic tangent activation function. The neural net has 9 input nodes (8 sensors and a bias), no hidden nodes and 2 output nodes (the left and right motor values for the differential drive), giving a total of 18 weights. To evolve these 18 weights, the evolutionary algorithm uses the obvious representation of real-valued vectors of length 18 for the genomes.

The robots' task is movement with obstacle avoidance: they have to learn to move around in a constrained arena with numerous obstacles as fast as possible while avoiding the obstacles. The robots are positioned in an arena with a small loop and varied non-looping corridors (see Fig. 1). The fitness function we use has been adapted from [14]; it favours robots that are fast and go straight ahead. Fitness is calculated as follows:

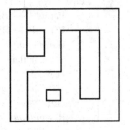

$$f = \sum_{t=0}^{\tau} (v_t \cdot (1 - v_r)) \tag{1}$$

where v_t and v_r are the translational and the rotational speed, respectively. v_t is normalised between -1 (full speed reverse) and 1 (full speed forward), v_r between 0 (movement in a straight line) and 1 (maximum rotation). In our simulations, whenever a robot touches an obstacle, $v_t = 0$, so the fitness increment for time-steps where the robot is in collision is 0. A

Fig. 1. The arena

good controller will turn only when necessary to avoid collisions and try to find paths that allow it to run in a straight line for as long as possible.

We run our simulations with each robot in an arena of its own so that they can't get in each others' way, although the robots obviously can communicate across arena instances. The reasons for this are twofold: firstly, eliminating physical interaction between the robots ensures that a robot's performance is due to its own actions rather than that of others around it; this allows us a clearer view of the effects of genome exchange. Secondly, it allows us to scale our simulations from a very small number of robots to a very large number of robots while using the exact same arenas; this guarantees that in those cases any change in performance of the robots is due to their increased group size rather than a change in environment.

We evaluate the EVAG variants with group sizes of 4, 16, 36 and 400 robots: we hypothesise that differences in group size influence the performance of distributed and hybrid algorithms due to their facilities for genome exchange. Since a larger group of robots is able to evaluate a larger number of candidate solutions simultaneously, the odds of finding a successful genome are higher. EVAG is intended to distribute these successful genomes across all robots, thus improving the performance of the entire group.

Because the experiments were designed so that genome exchange is the only possible interaction between robots in a group, the robots can have no physical interaction. Without physical interaction and no genome exchange the *average* performance of a group of 4 robots running $(\mu + 1)$ ON-LINE would be identical to that of a group of 400 robots doing the same. Since average performance is our only metric for success and since for $(\mu + 1)$ ON-LINE this metric is not influenced by the number of robots in the experiment, we can perform experiments for $(\mu + 1)$ ON-LINE for a group of 4 robots and use these results as a fair comparison to an experiment with EVAG using 400 robots. We can therefore safely omit the costly $(\mu + 1)$ ON-LINE simulations for the group sizes 16, 36 and 400 as their performance would be the same as that of the group of 4.[2]

Rather than comparing the algorithms at identical (so far as possible) parameter settings, we compare the performance at their relative *best possible* parameter settings after tuning as described in Sec 4.1. Note that, where applicable, values of μ may vary from one experiment to the next: this does not imply that the number of evaluations is influenced as the robots have a fixed amount of (simulated) wall-clock time in which to learn the task.

4.1 Evaluation with Parameter Tuning

It is a well-known fact that the performance of an evolutionary algorithm is greatly determined by its parameter values. Despite this, many publications in the field of evolutionary computing evaluate their algorithms using fairly arbitrary parameter settings, based on ad hoc choices, conventions, or a limited comparison of different parameter values. This approach can easily lead to misleading comparisons, where method A is tested with very good settings and method B based on poor ones. An recommendable alternative is the use of *automated parameter tuning*, where a tuning algorithm is used to optimize the parameter values and one compares the best variants of the evolutionary algorithms in question [3]. This approach helps prevent misleading comparions.

In our experiments with 4, 16 and 36 robots we evaluate the performance of the algorithms by performing parameter tuning for a fixed length of time and comparing the results. We use the MOBAT toolkit[3] to automate our tuning process. MOBAT is based on REVAC [10], which has been shown to be an efficient algorithm for parameter tuning [18]. For every combination of robot group size and algorithm MOBAT evaluates 400 parameter settings; each parameter setting is tested 25 times to allow statistically significant comparisons. Unless otherwise specified, performance comparisons were made with the best-performing parameter setting that was found for each of the algorithms-group size combinations.

Due to the computational requirements of tuning the parameters of very large simulations it was infeasible to tune parameters for our experiments with a group size of 400 robots. Instead, these experiments were performed at the parameter settings found for the 36 robots.

Each algorithm variant has its own parameters that need tuning. These parameters and the range within which they were tuned are listed in Table 1.

---

[2] Source code and scripts to repeat our experiments can be found at
http://www.few.vu.nl/~ehaasdi/papers/EA-2011-EvAg
[3] http://sourceforge.net/projects/mobat/

Table 1. The parameters as tuned by REVAC

Parameter	Tunable range
τ (Evaluation time steps per candidate)	300 – 600
μ (Population size – for encapsulated and hybrid schemes)	3 – 15
$\sigma_{initial}$ (Initial mutation step size)	0.1 – 10.0
ρ (Re-evaluation rate)	0.0 – 1.0
Crossover rate	0.0 – 1.0
Mutation rate	0.0 – 1.0
Newscast item TTL (for newscast-based EVAG variants)	3 – 20
Newscast cache size (for newscast-based EVAG variants)	2 – group size

5   Results and Discussion

Fig. 2 shows the results for the experiments for group sizes 4, 16, 36 and 400. Each graph shows the results for $(\mu + 1)$ ON-LINE (*encapsulated*), for EVAG (*distributed*) and for EVAG's *hybrid* extension. The latter two have results for the panmictic as well as the newscast variant. The white circles indicate the average performance (over 25 repeats) of the best parameter vector for that particular algorithm variant and the whiskers extend to the 95% confidence interval using a t-test. To investigate how sensitive the algorithm variants are to the choice of parameter settings, we also show the performance variation in the top 5% of parameter vectors, indicated by the grey ovals. The results are normalised so that the highest performance attained overall is 1. Note that because –as discussed above– we only ran $(\mu + 1)$ ON-LINE experiments with group size 4, the data for $(\mu + 1)$ ON-LINE are the same in all four graphs.

The performances shown are always the average performance of the entire group of robots in an experiment, not just the performance of the best robot: we are interested in developing an algorithm that performs well for *all* robots, rather than an algorithm that has a very high peak performance in one robot but does not succeed in attaining good performance for the entire group. Performances are compared using a t-test with $\alpha = 0.05$.

In the scenario with four robots (Fig. 2(a)) the hybrid algorithm significantly outperforms the encapsulated scheme, but it is not significantly better than the distributed scheme. The differences between the distributed and the encapsulated scheme are not significant. The same goes for the differences between the panmictic and the newscast variants. All four EVAG variants show a relatively large variation in the top-5% performances, indicating that they are more sensitive to the quality of parameter settings than is $(\mu + 1)$ ON-LINE.

It is surprising that the distributed scheme matches (even improves, although not significantly) performance with the encapsulated scheme when we consider that four robots running EVAG have access to only four (shared) genotypes, compared to the 12 (isolated) genotypes that are stored by each of the robots running $(\mu+1)$ ON-LINE (with μ tuned to 12).

With group size 16 (Fig. 2(b)), EVAG starts to come into its own: both the distributed and the hybrid schemes outperform $(\mu + 1)$ ON-LINE significantly, although the differences among the EVAG variants are not significant at 95%. Moreover, the distributed

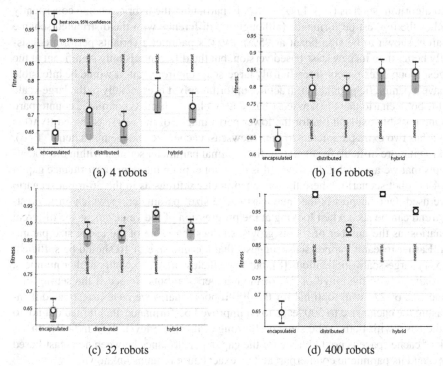

(a) 4 robots

(b) 16 robots

(c) 32 robots

(d) 400 robots

Fig. 2. Performance plots for various group sizes: performance is averaged over 25 repeats. Note that there was no tuning for group size 400 and hence no top 5% parameter vectors.

variants now perform almost as consistently as their encapsulated counterpart: both variation in scores for a single parameter setting and the variation of scores in the top-5% are at the level of $(\mu+1)$ ON-LINE, indicating that the distributed algorithm becomes less sensitive to sub-optimal parameter settings as the number of robots participating in the evolution grows.

With the increase in group size from 16 to 36 (Fig. 2(c)) EVAG again shows a significant performance increase and the confidence intervals and range of top-5% values are further reduced. For the first time we see a significant difference between the EVAG variants, with the panmictic hybrid approach performing significantly better than the alternatives.

Finally, Fig. 2(d) shows the results for 400 robots. The computational requirements of so large a simulation make parameter tuning infeasible; instead we investigate the performance of the algorithms at the best parameter settings found for the 36-robot scenario. The plot shows a large jump in performance for the panmictic variants, a respectable increase for the hybrid newscast variant, but little difference for the distributed newscast implementation. For all EVAG variants the confidence interval of the scores has shrunk considerably, indicating that EVAG continues to become more reliable as the robot group size increases.

The results show that, as the group size increases, there is an increasing benefit to using an algorithm such as (the hybrid extension of) EVAG rather than a purely encapsulated algorithm such as $(\mu + 1)$ ON-LINE. In particular, the hybrid scheme consistently reaches the highest performance (although the difference with the distributed scheme is rarely shown to be significant at 95%). EVAG's panmictic variants perform consistently better than its newscast-based version, but the difference is only significant at the largest group sizes we consider. In truly large-scale environments it would be infeasible to have a panmictic population structure; unfortunately it is especially in the large-scale 400-robot scenario that the newscast-based EVAG lags behind its panmictic counterpart.

One possible explanation for the lower performance of the newscast-based EVAG is that it has two extra parameters to tune (newscast cache size and the news items' TTL), making it more difficult for REVAC to find optimal parameter settings within the 400 attempts that we allowed it. However, this does not explain the large performance gap in the 400-robot scenario, where the same parameter settings as in the 36-robot scenario were used. One suspect is the 'newscast cache size' parameter, of which an interesting trend can be seen when looking at the progression of the value across the different scenarios: as the number of robots grows, so does the value of the cache size parameter. Earlier research on newscast suggests that a cache size of 10 should be sufficient for very large-scale applications [8], but nevertheless REVAC favours higher numbers, approximately in the range of $\frac{3}{4}$th of the number of robots. To see if the setting of a cache size of 27 is sub-optimal for the 400-robot scenario we have investigated if increasing the cache size to 300 leads to an improved performance; this turned out not to be the case, with both cache sizes performing at the same level. This indicates that a lack of cache space is not to blame for the gap in performance between newscast-based EVAG and its panmictic counterpart and its exact cause remains unknown.

6   Conclusion

In this paper we have compared the $(\mu + 1)$ ON-LINE on-board encapsulated algorithm to a distributed and a hybrid implementation of EVAG for on-line, on-board evolution of robot controllers. We have performed simulation experiments to investigate the performance of these algorithms, using automated parameter tuning to evaluate each algorithm at its *best possible* parameter setting.

Comparing the algorithms in terms of performance, EVAG performs consistently better than $(\mu + 1)$ ON-LINE, with the effect being especially prominent when the number of robots participating in the scenario is large. Even at the smallest group size we considered, the distributed scheme is competitive, despite having a population of only four individuals.

To estimate the sensitivity to sub-optimal parameter settings we have observed the variation in performance in the top-5% of parameter vectors. We have seen that $(\mu + 1)$ ON-LINE is quite stable, with the top-5% close to the best performance. In both the distributed and the hybrid variant, EVAG's performance is quite unstable when the group of robots is small, but becomes increasingly reliable as the group size increases.

For large numbers of robots, the newscast population structure has a small negative effect on the performance of the EVAG variants when compared to a panmictic population structure. However, in a large-scale scenario it may be infeasible to maintain a panmictic population structure. In those scenarios the small performance loss when using a newscast-based population structure may well be outweighed by the practical advantages of being able to implement the algorithm at all.

Although we have only performed experiments with a single and quite straightforward task, we conclude that both the distributed and hybrid approaches to on-board on-line evolutionary algorithms in evolutionary robotics are feasible and provide a promising direction for research in this field.

The hybrid scheme can be preferred over the encapsulated and the distributed case because it efficiently harnesses the opportunities of parallelising the adaptation process over multiple robots while performing well even for small numbers of robots. Although the panmictic variant does outperform the newscast-based implementation for very large numbers of robots, we do not know if or how tuning specifically for 400 robots would have influenced the apparent performance difference between newscast and panmixia. We would still prefer the latter because of its inherent scalability and robustness.

Future research should confirm our findings in different scenarios. Additionally, there is research to be done regarding the study of which parameters are influential and why certain parameter settings are more effective than others.

Acknowledgements. This work was made possible by the European Union FET Proactive Intiative: Pervasive Adaptation funding the SYMBRION project under grant agreement 216342. The authors would like to thank Selmar Smit and our partners in the SYMBRION consortium for many inspirational discussions on the topics presented here.

References

1. Bianco, R., Nolfi, S.: Toward open-ended evolutionary robotics: evolving elementary robotic units able to self-assemble and self-reproduce. Connection Science 4, 227–248 (2004)
2. Eiben, A.E., Karafotias, G., Haasdijk, E.: Self-adaptive mutation in on-line, on-board evolutionary robotics. In: 2010 Fourth IEEE International Conference on Self-Adaptive and Self-Organizing Systems Workshop (SASOW 2010), pp. 147–152. IEEE Press, Piscataway (2010)
3. Eiben, A.E., Smit, S.K.: Parameter tuning for configuring and analyzing evolutionary algorithms. Swarm and Evolutionary Computation 1(1), 19–31 (2011)
4. Elfwing, S., Uchibe, E., Doya, K., Christensen, H.I.: Biologically inspired embodied evolution of survival. In: Proceedings of the 2005 IEEE Congress on Evolutionary Computation, Edinburgh, UK, September 2-5, vol. 3, pp. 2210–2216. IEEE Press (2005)
5. Floreano, D., Schoeni, N., Caprari, G., Blynel, J.: Evolutionary bits'n'spikes. In: Standish, R.K., Bedau, M.A., Abbass, H.A. (eds.) Artificial Life VIII: Proceedings of the Eighth International Conference on Artificial Life, pp. 335–344. MIT Press, Cambridge (2002)
6. Giacobini, M., Preuss, M., Tomassini, M.: Effects of Scale-Free and Small-World Topologies on Binary Coded Self-adaptive CEA. In: Gottlieb, J., Raidl, G.R. (eds.) EvoCOP 2006. LNCS, vol. 3906, pp. 86–98. Springer, Heidelberg (2006)
7. Haasdijk, E., Eiben, A.E., Karafotias, G.: On-line evolution of robot controllers by an encapsulated evolution strategy. In: Proceedings of the 2010 IEEE Congress on Evolutionary Computation, Barcelona, Spain, IEEE Computational Intelligence Society, IEEE Press (2010)

8. Jelasity, M., van Steen, M.: Large-scale newscast computing on the internet. Technical report, Vrije Universiteit Amsterdam (2002)
9. Laredo, J.L., Eiben, A.E., van Steen, M., Merelo, J.J.: Evag: a scalable peer-to-peer evolutionary algorithm. Genetic Programming and Evolvable Machines 11, 227–246 (2010)
10. Nannen, V., Eiben, A.E.: Relevance estimation and value calibration of evolutionary algorithm parameters. In: Proc. of IJCAI 2007, pp. 975–980 (2007)
11. Nannen, V., Smit, S.K., Eiben, A.E.: Costs and Benefits of Tuning Parameters of Evolutionary Algorithms. In: Rudolph, G., Jansen, T., Lucas, S., Poloni, C., Beume, N. (eds.) PPSN X. LNCS, vol. 5199, pp. 528–538. Springer, Heidelberg (2008)
12. Nehmzow, U.: Physically embedded genetic algorithm learning in multi-robot scenarios: The pega algorithm. In: Prince, C.G., Demiris, Y., Marom, Y., Kozima, H., Balkenius, C. (eds.) Proceedings of The Second International Workshop on Epigenetic Robotics: Modeling Cognitive Development in Robotic Systems, Edinburgh, UK. Lund University Cognitive Studies, vol. 94. LUCS (August 2002)
13. Nelson, A.L., Barlow, G.J., Doitsidis, L.: Fitness functions in evolutionary robotics: A survey and analysis. Robotics and Autonomous Systems 57(4), 345–370 (2009)
14. Nolfi, S., Floreano, D.: Evolutionary Robotics: The Biology, Intelligence, and Technology of Self-Organizing Machines. MIT Press, Cambridge (2000)
15. Nordin, P., Banzhaf, W.: Genetic programming controlling a miniature robot. In: Working Notes for the AAAI Symposium on Genetic Programming, pp. 61–67. AAAI (1995)
16. Nordin, P., Banzhaf, W.: An on-line method to evolve behavior and to control a miniature robot in real time with genetic programming. Adaptive Behavior 5, 107–140 (1997)
17. Schwefel, H.-P.: Numerical Optimization of Computer Models. John Wiley & Sons, Inc., New York (1981)
18. Smit, S.K., Eiben, A.E.: Comparing parameter tuning methods for evolutionary algorithms. In: Proceedings of the Eleventh Conference on Congress on Evolutionary Computation, CEC 2009, Piscataway, NJ, USA, pp. 399–406. IEEE Press (2009)
19. Usui, Y., Arita, T.: Situated and embodied evolution in collective evolutionary robotics. In: Proceedings of the 8th International Symposium on Artificial Life and Robotics, pp. 212–215 (2003)
20. Walker, J.H., Garrett, S.M., Wilson, M.S.: The balance between initial training and lifelong adaptation in evolving robot controllers. IEEE Transactions on Systems, Man, and Cybernetics, Part B 36(2), 423–432 (2006)
21. Watson, R.A., Ficici, S.G., Pollack, J.B.: Embodied evolution: Distributing an evolutionary algorithm in a population of robots. Robotics and Autonomous Systems 39(1), 1–18 (2002)

Improving Performance via Population Growth and Local Search: The Case of the Artificial Bee Colony Algorithm

Doğan Aydın[1], Tianjun Liao[2], Marco A. Montes de Oca[3], and Thomas Stützle[2]

[1] Dept. of Computer Engineering, Dumlupınar University, 43030 Kütahya, Turkey
dogan.aydin@dpu.edu.tr
[2] IRIDIA, CoDE, Université Libre de Bruxelles, Brussels, Belgium
{tliao,stuetzle}@ulb.ac.be
[3] Dept. of Mathematical Sciences, University of Delaware, Newark, DE, USA
mmontes@math.udel.edu

Abstract. We modify an artificial bee colony algorithm as follows: we make the population size grow over time and apply local search on strategically selected solutions. The modified algorithm obtains very good results on a set of large-scale continuous optimization benchmark problems. This is not the first time we see that the two aforementioned modifications make an initially non-competitive algorithm obtain state-of-the-art results. In previous work, we have shown that the same modifications substantially improve the performance of particle swarm optimization and ant colony optimization algorithms. Altogether, these results suggest that population growth coupled with local search help obtain high-quality results.

1 Introduction

Thinking of optimization algorithms as being composed of components has changed the way high-performance optimization algorithms are designed. Research based on algorithm components and not on specific implementations of optimization algorithms promises to lead to breakthroughs because a more systematic approach can be taken in order to explore the space of algorithm designs. In previous work, we have taken a component-based approach to the design of optimization algorithms for continuous optimization and have obtained promising results. In particular, we integrated a growing population size and a local search components into a particle swarm optimization (PSO) algorithm and an ant colony optimization algorithm for continuous domains. The resulting algorithms, called IPSOLS [3,4] in the case of PSO, and IACO$_\mathbb{R}$-LS [5] in the case of ACO$_\mathbb{R}$, exhibited a significantly better performance than the original algorithms. In fact, their performance was competitive with state-of-the-art algorithms in the special issue of the Soft Computing journal on large-scale continuous optimization [6]. (Throughout the rest of the paper, we will refer to this special issue as SOCO.) In this paper, we present a third case study based on the artificial

J.-K. Hao et al. (Eds.): EA 2011, LNCS 7401, pp. 85–96, 2012.
© Springer-Verlag Berlin Heidelberg 2012

bee colony (ABC) algorithm [7, 8]. The goal of this case study is to test the "pure chance" hypothesis, which explains the results obtained with PSO and $ACO_{\mathbb{R}}$ as consequence of mere good luck. The results obtained in this third case study, shown in Section 3.4, suggest that the pure-chance hypothesis is false. We find that population growth together with a local search procedure are algorithmic components that make swarm intelligence algorithms for continuous optimization obtain high-quality results.

2   Related Work

The idea of increasing the size of the population in swarm intelligence algorithms for continuous optimization derives from the incremental social learning (ISL) framework [3]. The ISL framework's aim is to reduce the time needed by a swarm intelligence system to reach a desired state. In the case of swarm intelligence systems for optimization, the desired state may be associated with a solution better than or equal to a desired quality. The ISL framework achieves this goal by starting the system with a small population and increasing its size according to some addition criterion. By starting a swarm-based optimization algorithm with a smaller population than usual, the framework encourages a rapid convergence toward promising regions of the search space. Adding new solutions increases the diversity of the swarm. The new solutions are generated using information from the "experienced" swarm; thus, new solutions are not completely random.

Modifying the size of the population while an algorithm is operating is not a new idea. For example, a number of researchers in the field of evolutionary computation (EC) have proposed schemes that increase or reduce the size of the population in order to adjust the diversification-intensification properties of the underlying optimization algorithms (see e.g. [9–11]). The ISL framework is different from most previous dynamic population sizing approaches. In most of these cases, diversification is favored during the first phases of the optimization process through a large population, and intensification toward the end by reducing its size. In contrast, the ISL framework modifies the population size in one direction only: increasing its size. A notable exception of the strategy used in most EC algorithms is the one used in the G-CMA-ES [12] algorithm. In this algorithm, the population size also increases over time. It is important to note that G-CMA-ES is considered to be a state-of-the-art algorithm for continuous optimization.

The utilization of an auxiliary local search method is a common approach to enhance the intensification properties of EC algorithms. In ISL, local search can be integrated as a means to simulate individual learning, that is, learning without any social influence.

In [3], we introduced IPSOLS, a PSO-local search hybrid algorithm with increasing population size as an instantiation of the ISL framework. In subsequent work [4], we used iterated F-Race [14], an automatic algorithm configuration software, throughout the redesign process of IPSOLS. The redesigned IPSOLS algorithm [4] was benchmarked on SOCO's large-scale benchmark problems and compared favorably to other state-of-the-art algorithms that include differential evolution algorithms, memetic algorithms, particle swarm optimization algorithms and other types of optimization algorithms [6]. In SOCO, a differential evolution algorithm (DE) [15], G-CMA-ES, and the real coded CHC algorithm (CHC) [16] were used as reference algorithms. A second instantiation of ISL was presented in [5] in the context of ACO$_\mathbb{R}$. The introduced algorithm, called IACO$_\mathbb{R}$-LS, features the same two components: a growing solution population size and a local search procedure. IACO$_\mathbb{R}$-LS was also benchmarked on SOCO's benchmark problems and on the IEEE CEC 2005 functions [17]. We can say that IACO$_\mathbb{R}$-LS is a state-of-the-art algorithm because it obtained better results than IPSOLS and it is competitive with G-CMA-ES.

3 An Artificial Bee Colony Algorithm with Population Growth and Local Search

In this section, we present our third case study of the utilization of a growing population size and a local search procedure in a swarm intelligence algorithm for continuous optimization.

3.1 The Artificial Bee Colony Algorithm

The ABC algorithm [7, 8] is inspired by the foraging behavior of a honeybee swarm. At the initialization step of the algorithm, ABC generates a number of randomly located food sources, and it creates a number of employed and onlooker bees. The number of food sources, which is denoted by SN, is equal to the number of employed and onlooker bees. Each cycle of the algorithm consists of three successive steps. In the first step, each employed bee selects successively a food source i and then produces a candidate food, v_i, by mutating the location of the selected food source as reference according to

$$v_{i,j} = x_{i,j} + \phi_{i,j}(x_{i,j} - x_{k,j}), \, i \neq k, \tag{1}$$

where $k \in \{1, 2, \ldots, SN\}$, $j \in \{1, 2, \ldots, D\}$ (D is the problem's dimension), $\phi_{i,j}$ is a uniform random number in $[-1, 1]$, $x_{i,j}$ and $x_{k,j}$ is the position of the reference food source i and a randomly selected food source k in dimension j. The candidate food source created by an employed bee can be better than the reference food source. In this case, the reference food source, x_i, is replaced with the candidate food source, v_i. In the second step, onlooker bees try to find new food sources near existing food sources as the employed bees do. Different from employed bees, onlooker bees randomly select a food source i with a food source selection probability p_i, which is determined as

$$p_i = \frac{fitness_i}{\sum_{n=1}^{SN} fitness_n}, \tag{2}$$

where $fitness_i$ is the $fitness$ value of the food source i, which is inversely proportional to the objective value of the food source i for function minimization.

Algorithm 1. The Incremental ABC Algorithm with Local Search

---

initialization {Initialize SN food sources}
while termination condition is not met **do**
    if $FailedAttempts = Failures_{Max}$ **then**
        Invoke local search on randomly selected food source
    else
        Invoke local search on the best food source
    end if
    Employed Bees Stage {Use $x_{gbest,j}$ as the reference food source}
    Onlookers Stage
    Scout Bees Stage {Use Eq. 5}
    if Food source addition criterion is met **then**
        Add a new food source to the environment {Use Eq. 4}
        $SN \leftarrow SN + 1$
    end if
end while

Onlooker bees explore in the vicinity of good food sources. This behavior is responsible for the intensification behavior of the algorithm since information about good solutions is exploited. In the last step, a few food sources, which have not been improved during a predetermined number of iterations (controlled by a parameter *limit*), are detected and abandoned. Then, scout bees search randomly for a new food source, x_i, and replaces with it the location of the abandoned food source, x_i^{old}. The new food source is found according to

$$x_{i,j} = x_j^{min} + \varphi_{i,j}(x_j^{max} - x_j^{min}) \tag{3}$$

where $\varphi_{i,j}$ is a uniform random number in $[0,1]$ for dimension j and food source i, and x_j^{min} and x_j^{max} are the minimum and maximum limits of the search range on dimension j, respectively. Clearly, in the ABC algorithm employed and onlooker bees intensify the algorithm's search and the scout bees diversify it [18].

3.2 Integrating Population Growth and Local Search

In this section, we describe the integration of a growing population size and a local search procedure into the ABC algorithm. The outline of the proposed algorithm, called IABC-LS, is shown in Algorithm 1. In IABC-LS, the number of food sources and, indirectly, the population size of the bee colony (that is, the number of onlooker and employed bees), is increased according to a control parameter g. Every g iterations, a new food source is added to the environment until a maximum number of food sources is reached.

IABC-LS begins with few food sources. New food sources are placed biasing their location toward the location of the best-so-far solution. This is implemented as

$$x'_{new,j} = x_{new,j} + \varphi_{new,j}(x_{gbest,j} - x_{new,j}), \tag{4}$$

where $x_{new,j}$ is the randomly generated new food source location in dimension j, $x'_{new,j}$ is the updated location of the new food source, $x_{gbest,j}$ refers to best-so-far food source location, and $\varphi_{new,j}$ is a number chosen uniformly at random in $[0,1]$. A similar replacement mechanism is applied by the scout bees step in IABC-LS. The difference is a replacement factor, R_{factor}, that controls how much the new food source locations will be closer to the best-so-far food source. This modified rule is:

$$x'_{new,j} = x_{gbest,j} + R_{factor}(x_{gbest,j} - x_{new,j}). \tag{5}$$

Another difference between the original ABC algorithm and IABC-LS is that employed bees search in the vicinity of $x_{gbest,j}$ instead of a randomly selected food source. This modification enhances the intensification behavior of the algorithm and helps it to converge quickly toward good solutions. Although intensification in the vicinity of the best-so-far solution may lead to premature convergence, the algorithm detects stagnation by using the *limit* parameter, and the scouts can discover another random food source to avoid search stagnation.

IABC-LS is a hybrid algorithm that calls a local search procedure at each iteration. The best-so-far food source location is usually used as the initial solution from which the local search is called. The result of the local search replaces the best-so-far solution if there is an improvement on the initial solution. In this paper, the IABC-LS algorithm is implemented with Powell's conjugate direction set [19] (IABC-Powell) and Lin-Yu Tseng's Mtsls1 [20] (IABC-Mtsls1) as local search procedures. Both local search procedures are terminated after a maximum number of iterations, itr_{max}, or when the tolerance $FTol$ is reached. $FTol$ is the threshold value of the relative difference between two successive iterations' solutions. An adaptive step size for each local search procedure is used. The step size is set to the maximum norm ($\| \cdot \|_\infty$) of the vector that separates a randomly selected food source from the best-so-far food source.

For fighting stagnation, the local search procedures are applied to a randomly selected location if local search calls cannot improve the results after a maximum number of repeated calls, which is controlled by a parameter $Failures_{max}$. The original versions of the local search algorithms do not enforce bound constraints. To enforce bound constraints, the following penalty function is used in both local search procedures [4,5]:

$$P(x) = fes \times \sum_{j=1}^{D} Bound(x_j), \tag{6}$$

where $Bound(x_j)$ is defined as

$$Bound(x_j) = \begin{cases} 0, & \text{if } x_j^{max} > x_j > x_j^{min} \\ (x_j^{min} - x_j)^2, & \text{if } x_j < x_j^{min} \\ (x_j^{max} - x_j)^2, & \text{if } x_j > x_j^{max} \end{cases} \tag{7}$$

where *fes* is the number of function evaluations that have been used so far.

Table 1. The best parameter configurations obtained through iterated F-race for ABC algorithms in 10-dimensional instances

Algorithm	$SN\ limit$	SN_{max}	$growth$	R_{factor}	itr_{max}	$Failures_{max}$	$FTol$
ABC	5	98	—	—	—	—	—
IABC	8	96	13	8	10^{-6}	—	—
IABC-Mtsls1	4	13	80	6	$10^{-2.94}$	199	17	—
IABC-Powell	7	99	44	9	$10^{-0.94}$	82	9	$10^{-5.6}$

3.3 Experimental Setup

In our study, we performed three sets of experiments. First, we ran the original ABC algorithm, IABC (IABC-LS without local search) and the two instantiations of IABC-LS on the 19 SOCO benchmark functions proposed by Herrera et al. [21] for SOCO. A detailed description of the benchmark function set can be found in [21]. All experiments were conducted under the same conditions, and grouped by the dimensionality of the optimization problems ($D \in \{50, 100, 200, 500, 1000\}$). All investigated algorithms were run 25 times for each SOCO function. Each run stops when the maximum number of evaluations is achieved or the solution value is lower than 10^{-14}, which is approximated to 0. The maximum number of evaluations is $5000 \times D$.

The parameters of all ABC algorithms used in this paper were tuned using iterated F-race [14]. The best parameters found for each algorithm are given in Table 1. Iterated F-race was run using the 10-dimensional versions of the 19 benchmark functions as tuning instances and the maximum tuning budget was set to 50,000 algorithm runs. The number of function evaluations used in each run is 50,000.

We compared the computational results of IABC-Powell and IABC-Mtsls1 with IACO$_\mathbb{R}$-LS and IPSOLS. Finally, all three IABC variants were compared with the 16 algorithms featured in the SOCO special issue.

3.4 Results

The results of the first set of experiments, where we compare the ABC algorithm variants, are shown in Figure 1. The box-plots indicate the distribution of average results (left side) and the median results (right side) obtained by each algorithm on the 19 SOCO benchmark functions. (Each point for the box-plot measures the mean and median performance for 25 independent trials on each function.) In all cases, except in the comparison between IABC-Mtsls1 and ABC on the 50-dimensional functions, the mean and median results obtained by the proposed algorithms are statistically significantly better than those of the original ABC algorithm. Statistical significance was determined with Wilcoxon's test at $\alpha = 0.05$ using Holm's method for multiple test corrections. There are interesting differences in performance depending on whether we base our conclusions

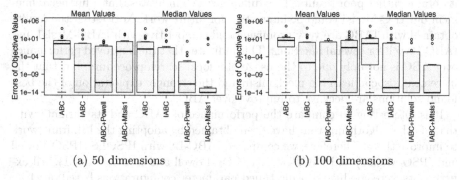

(a) 50 dimensions   (b) 100 dimensions

Fig. 1. Distribution of mean and median errors of ABC, IABC, IABC-Powell and IABC-Mtsls1 on the 19 SOCO benchmark functions

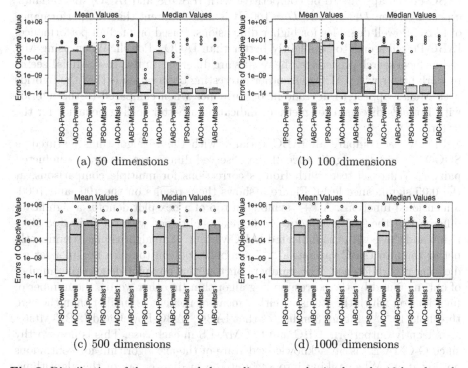

(a) 50 dimensions   (b) 100 dimensions

(c) 500 dimensions   (d) 1000 dimensions

Fig. 2. Distribution of the mean and the median errors obtained on the 19 benchmark functions

on the median or the mean performance. Based on mean errors, IABC-Powell performs better than IABC and IABC-Mtsls1. However, based on median errors (that is, for each function, we measure the median result across the 25 independent trials), IABC-Mtsls1 outperforms all others. This difference is the result of the fact that IABC-Mtsls1 shows early stagnation effects in few trials in which it obtains rather poor results. This difference also indicates that a further refinement of IABC-Mtsls1 could be interesting for future work. Notice that the results obtained with IABC are very promising and competitive with IABC-Mtsls1 and IABC-Powell for several functions. This indicates that the improved performance over ABC is not only due to the usage of a local search procedure, but that the incremental social learning mechanism and the stronger exploitation of the best found solutions are decisive to improve over ABC.

It is interesting to compare the performance of IABC and its variants with that of other algorithms that have been obtained by adopting the ISL framework to improve them. Therefore, we compared IABC-LS with IPSOLS (IPSO-Powell and IPSO-Mtsls1) and IACO$_\mathbb{R}$-LS (IACO$_\mathbb{R}$-Powell and IACO$_\mathbb{R}$-Mtsls1). All experiments were conducted using tuned parameter configurations based on [4,5]. The distributions of the mean and median errors obtained on the 19 SOCO test functions for various dimensions are shown in Figure 2. Based on the mean errors, IABC-Powell appears to be competitive with IPSOLS and IACO$_\mathbb{R}$-LS variants, especially when $D = 50$ and $D = 100$. It is also apparent that the performance of IABC-Powell degrades for higher dimensions. Based on median performance, IABC-Powell seems to be slightly better than IACO$_\mathbb{R}$-Powell. When using Mtsls1 as local search procedure, the results are different. In this case, Mtsls1 seems to work better with IACO$_\mathbb{R}$. Another interesting observation is that the hybrids with Mtsls1 appear to degrade in performance when compared to the hybrids with Powell's direction set method, indicating a better scaling behavior for the latter.

Finally, we compare all IABC variants with the 16 algorithms featured in SOCO. To test the significance of the observed differences, we again conducted pairwise Wilcoxon tests with Holm's corrections for multiple comparisons, at the 0.05 significance level. Figure 3 shows these results on the 100- and 1000-dimensional functions. A "+" symbol on top of a box-plot denotes a significant difference at the 0.05 level between the results obtained with the indicated algorithm and those with the indicated IABC variant. If an algorithm is significantly better than an IABC variant, a "−" symbol is put on top of a box-plot for indicating this difference. The numbers on top of a box-plot denote the number of optima found by the corresponding algorithm (in other words, the number of functions on which the statistic, either mean or median, is smaller than the zero threshold 10^{-14}). Figure 3 indicates that IABC, IABC-Powell and IABC-Mtsls1 significantly outperform CHC and G-CMA-ES in each case. This is noteworthy since G-CMA-ES is an acknowledged state-of-the-art algorithm for continuous optimization. When taking into account the median errors of the algorithms, IABC variants outperform CHC, G-CMA-ES, EvoPROpt, MA-SSW, RPSO-vm, and VXQR1 in almost all dimensions. The IABC variants exhibit a performance similar to rest of the state-of-the-art algorithms with the exception of IPSO-Powell and MOS-DE.

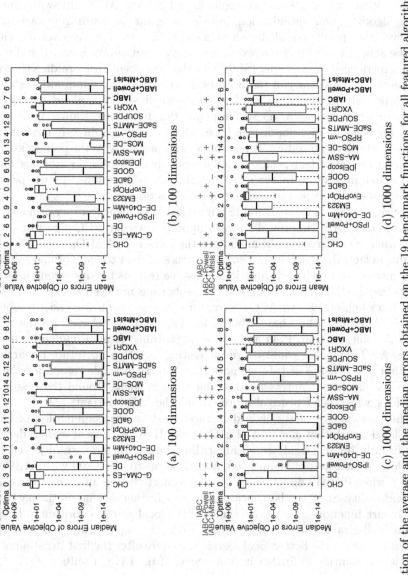

Fig. 3. Distribution of the average and the median errors obtained on the 19 benchmark functions for all featured algorithms in SOCO and IABC variants on 100- and 1000-dimensional functions. G-CMA-ES results on 1000-dimensional functions are unknown.

4   Discussion

In this paper, we have shown that the ISL framework with local search can enhance the performance of ABC. In earlier work we have shown the same result for two other swarm intelligence algorithms: PSO and $ACO_{\mathbb{R}}$. In swarm intelligence algorithms for optimization, individual agents can learn from each other. However, the specific knowledge-sharing mechanism used in any given algorithm can be slow. In the ISL framework, new agents can obtain knowledge directly from experienced ones. Thus, ISL allows the new agent to more directly focus the search in the vicinity of the best results. On the other hand, although local search procedures cannot always find good-enough solutions alone, local search procedures play an important role in population-based algorithms. At the same time, the behavior of a population-based algorithm affects directly the behavior of the local search procedure. For example, information from the population may be used to select restart positions and adaptively determine the step size of the local search procedure. In our case, the step size is determined by the position of the best-so-far agent and another agent's position.

Concerning the impact of the ISL framework on algorithm performance, we observed that IABC improved strongly upon the original ABC algorithm, more than what we observed when moving to the incremental $ACO_{\mathbb{R}}$ and PSO algorithms. In fact, the IABC algorithm alone was sufficient to find optimal solutions for a number of benchmark functions. Possible reasons for this strong improvements may be that (i) the bee colony metaphor has not been as explored as the metaphors behind PSO or $ACO_{\mathbb{R}}$ have; thus, improvements are easier to obtain; and (ii) ABC algorithms have a better local search type behavior for the refinement of solutions than PSO and $ACO_{\mathbb{R}}$ algorithms. A more in-depth analysis of the ABC algorithm could be an interesting direction for future research. In addition to the usage of ISL, we note that another possible reason for the observed performance improvements is the stronger focus around the best-so-far solutions, which is a feature also present in other ABC variants [22].

The performance of the hybrid method, that is, the combination of IABC with local search (but also that of $IACO_{\mathbb{R}}$ and IPSO with local search) depends significantly on the local search algorithm chosen. In fact, we observed that for "low" dimensions of 50 and 100, Mtsls1 seems to work well with IABC and $ACO_{\mathbb{R}}$, while Powell's direction set method seems to be less affected by increasing dimensionality. Since the best local search may further depend on the particular benchmark function, an adaptive choice of the local search algorithm to be used would be desirable. As example, we have tried a simple strategy in which the algorithm selects the better local search procedure after the first iteration of the algorithm, resulting in further improvements of the IABC results.

5   Conclusions

In this paper, we have introduced an ABC algorithm with a growing population size and we hybridized it with a local search procedure for tackling large-scale benchmark functions. The increasing population size and a stronger focus on the best-so-far solution have contributed to make the proposed IABC algorithm much better performing than the original ABC algorithm. For the hybridization with local search two different local search procedures were used, Powell's conjugate direction set and Mtsls1. The parameters of IABC and the hybrid IABC with local search were tuned using iterated F-Race, an automatic algorithm configuration tool. For the benchmark functions of 50 and 100 dimensions we found that the hybrid algorithm typically improves over IABC. When compared with other algorithms for large scale continuous optimization from the SOCO special issue, the hybrid IABC algorithm was found to reach high-quality solutions; it was outperformed, according to the median results, only by an incremental PSO algorithm and the overall best performing algorithm from the SOCO benchmark competition.

It is maybe more noteworthy that also in the ABC case, extending this algorithm with the incremental social learning framework led to significant performance improvements. Certainly, further analysis of the hybrid IABC algorithm is necessary to determine the contribution of the specific algorithm components. Nevertheless, the fact that apart from ABC we also could improve PSO and continuous ACO algorithms by embedding them into the incremental social learning framework, gives strong evidence that an increasing population size and the hybridization with local search algorithms are important to obtain high performing swarm intelligence algorithms for continuous optimization.

Acknowledgments. The research leading to the results presented in this paper has received funding from the European Research Council under the European Union's Seventh Framework Programme (FP7/2007-2013) / ERC grant agreement n°246939, and by the Meta-X project, funded by the Scientific Research Directorate of the French Community of Belgium. Thomas Stützle acknowledges support from the Belgian F.R.S.-FNRS, of which he is a Research Associate.

References

1. Kennedy, J., Eberhart, R.: Particle swarm optimization. In: Proc. IEEE International Conference of Neural Networks, vol. 4, pp. 1942–1948 (1995)
2. Socha, K., Dorigo, M.: Ant colony optimization for continuous domains. European Journal of Operational Research 185, 1155–1173 (2008)
3. Montes de Oca, M.A., Stützle, T., Van den Enden, K., Dorigo, M.: Incremental social learning in particle swarms. IEEE Transactions on Systems, Man, and Cybernetics, Part B: Cybernetics 41(2), 368–384 (2011)
4. Montes de Oca, M.A., Aydın, D., Stützle, T.: An incremental particle swarm for large-scale optimization problems: An example of tuning-in-the-loop (re)design of optimization algorithms. Soft Computing 15(11), 2233–2255 (2011)
5. Liao, T., Montes de Oca, M.A., Aydın, D., Stützle, T., Dorigo, M.: An Incremental Ant Colony Algorithm with Local Search for Continuous Optimization. In: GECCO 2011, pp. 125–132. ACM Press, New York (2011)

6. Lozano, M., Molina, D., Herrera, F.: Editorial: Scalability of evolutionary algorithms and other metaheuristics for large-scale continuous optimization problems. Soft Computing 15(11), 2085–2087 (2011)
7. Karaboga, D.: An idea based on honey bee swarm for numerical optimization. Technical Report-TR06, Erciyes Universitesi, Computer Engineering Department (2005)
8. Karaboga, D., Basturk, B.: A powerful and efficient algorithm for numerical function optimization: artificial bee colony (ABC) algorithm. Journal of Global Optimization 39(3), 459–471 (2007)
9. Eiben, A.E., Marchiori, E., Valkó, V.A.: Evolutionary Algorithms with On-the-Fly Population Size Adjustment. In: Yao, X., Burke, E.K., Lozano, J.A., Smith, J., Merelo-Guervós, J.J., Bullinaria, J.A., Rowe, J.E., Tiňo, P., Kabán, A., Schwefel, H.-P. (eds.) PPSN VIII. LNCS, vol. 3242, pp. 41–50. Springer, Heidelberg (2004)
10. Chen, D., Zhao, C.: Particle swarm optimization with adaptive population size and its application. Applied Soft Computing 9(1), 39–48 (2009)
11. Hsieh, S., Sun, T., Liu, C., Tsai, S.: Efficient population utilization strategy for particle swarm optimizer. IEEE Transactions on Systems, Man, and Cybernetics, Part B: Cybernetics 39(2), 444–456 (2009)
12. Auger, A., Hansen, N.: A restart CMA evolution strategy with increasing population size. In: CEC 2005, pp. 1769–1776. IEEE Press (2005)
13. Balaprakash, P., Birattari, M., Stützle, T.: Improvement Strategies for the F-Race Algorithm: Sampling Design and Iterative Refinement. In: Bartz-Beielstein, T., Blesa Aguilera, M.J., Blum, C., Naujoks, B., Roli, A., Rudolph, G., Sampels, M. (eds.) HM 2007. LNCS, vol. 4771, pp. 108–122. Springer, Heidelberg (2007)
14. Birattari, M., Yuan, Z., Balaprakash, P., Stützle, T.: F-Race and iterated F-Race: An overview. In: Experimental Methods for the Analysis of Optimization Algorithms, pp. 311–336. Springer, Heidelberg (2010)
15. Storn, R., Price, K.: Differential evolution–a simple and efficient heuristic for global optimization over continuous spaces. Journal of Global Optimization 11(4), 341–359 (1997)
16. Eshelman, L., Schaffer, J.: Real-coded genetic algorithms and interval-schemata. Foundations of Genetic Algorithms 2, 187–202 (1993)
17. Suganthan, P., Hansen, N., Liang, J., Deb, K., Chen, Y., Auger, A., Tiwari, S.: Problem definitions and evaluation criteria for the CEC 2005 special session on real-parameter optimization. Technical Report 2005005, Nanyang Technological University (2005)
18. Karaboga, D., Akay, B.: A comparative study of Artificial Bee Colony algorithm. Applied Mathematics and Computation 214(1), 108–132 (2009)
19. Powell, M.: An efficient method for finding the minimum of a function of several variables without calculating derivatives. The Computer Journal 7(2), 155–162 (1964)
20. Tseng, L., Chen, C.: Multiple trajectory search for large scale global optimization. In: CEC 2008, pp. 3052–3059. IEEE Press, Piscataway (2008)
21. Herrara, F., Lozano, M., Molina, D.: Test suite for the special issue of soft computing on scalability of evolutionary algorithms and other metaheuristics for large scale continuous optimization problems (2010),
http://sci2s.ugr.es/eamhco/updated-functions1-19.pdf
22. Diwold, K., Aderhold, A., Scheidler, A., Middendorf, M.: Performance evaluation of artificial bee colony optimization and new selection schemes. Memetic Computing 3(3), 149–162 (2011)

Two Ports of a Full Evolutionary Algorithm onto GPGPU

Ogier Maitre, Nicolas Lachiche, and Pierre Collet

LSIIT, University of Strasbourg, 67412 Illkirch, France
{ogier.maitre,nicolas.lachiche,pierre.collet}@unistra.fr

Abstract. This paper presents two parallelizations of a standard evolutionary algorithm on an NVIDIA GPGPU card, thanks to a parallel replacement operator.

These algorithms tackle new problems where previously presented approaches do not obtain satisfactory speedup. If programming is more complicated and fewer options are allowed, the whole algorithm is executed in parallel, thereby fully exploiting the intrinsic parallelism of EAs and the many available GPGPU cores.

Finally, the method is validated using two benchmarks.

1 Introduction

The use of General Purpose Graphic Processing Unit (GPGPU) many-core processors for generic computing has raised a lot of interest within the scientific community. Artificial evolution is no exception to this trend and many works propose to port evolutionary algorithms onto GPGPU cards.

Many papers have been published on porting Genetic Programming on GPGPU, because of its great need for computing power. In addition, its model is highly asymmetric: evaluation is by far the most consuming part of the algorithm. Yet genetic algorithms and evolution strategies have proven to be effective and many problems could benefit from a massive parallelization of EAs on multicore processors, in order to widen search space exploration, or to find equivalent results faster and therefore allow EAs to tackle much larger problems.

Nevertheless, only few works try to port a standard EA on a GPU. When this is done, the algorithm is non standard and performance comparisons are therefore difficult.

In this paper, we propose two complete parallelizations of evolutionary algorithms on GPGPU, which are comparable with sequential algorithms. Thus, users familiar with evolutionary algorithms can access the GPGPU cards computing power without losing their habits. These implementations exhibit interesting speedups with various algorithm configurations.

2 Related Works

2.1 Parellization of Evolutionary Algorithms onto GPGPU

Different implementations of evolutionary algorithms on GPGPU have been attempted in the past. Genetic programming is part of the approaches discussed

J.-K. Hao et al. (Eds.): EA 2011, LNCS 7401, pp. 97–108, 2012.

most recently. This trend is surely due to the evaluation time needed to compute the fitness value of GP individuals. Among these works, we can cite [3], [8] and [11].

Other studies target genetic algorithms and evolutionary strategies. Works of this type are presented in [2,4,13], where the authors program the card at the vertex and pixel shader level. The use of such programming paradigms makes porting complex and requires algorithmic changes that tend to make the resulting algorithms difficult to compare with sequential ones. [4,13] use techniques such as neighbouring selection, while [2] uses a selection with replacement in order to select next parents, which could affect the diversity of the population (there may be clones in the new generation). Furthermore, the raw speedups of these approaches are disappointing considering the efforts invested, even if it is necessary to keep in mind the relative age of the equipment used.

Another approach presented in [10] uses an island model in which each processor group manages an island. The local population is stored in on-chip memory and implementation shows impressive speedups($\approx$ 7000$\times$). However the use of this memory limits the size of the individuals and the number of individuals into an island. Moreover the use of an internal island model on the card questions about the scalability of this approach to a multi-cards architecture.

Recently, the EASEA platform implements another model of parallel evolutionary algorithm on GPGPU. The evolutionary engine runs on the CPU and only evaluation is parallelized on GPU: the population is transferred to the GPU card to be evaluated, then only the fitness values are transferred back to CPU memory in order to continue the algorithm. This model is similar to a standard sequential algorithm, but as stated in Amdhal's law [1], the gain of porting an algorithm on a parallel architecture is limited by the sequential part of the algorithm. Furthermore, it is similar to what is done for the parallelization of GP on GPGPU: only costly evaluation functions can really benefit from this model. The EASEA plateform is thoroughly presented in [7].

2.2   DISPAR: A Parallel Remplacement Operator

One of the main problems about parallelizing a complete EA is the population reduction step. In the case of an (1000+500)ES algorithm, for instance, with 1000 individuals in the population and 500 children, it is necessary to reduce the temporary population of parents + children (1500 individuals) back to only 1000 individuals for the next generation.

Ideally, this reduction operator must not select twice the same individual, in order to preserve diversity in the new generation. The selection must be operated without replacement (an individual selected to be part of the new generation must be removed from the initial pool of individuals). This is very problematic if a parallel selection were to be performed, as it would be impossible to prevent two different cores from selecting the same individual.

DISPAR-tournament (Disjoint Sets Parallel tournament) is a parallel selection operator that mimics a tournament without replacement [9]. The population is divided into blocks of T neighbouring individuals, each block being given to a

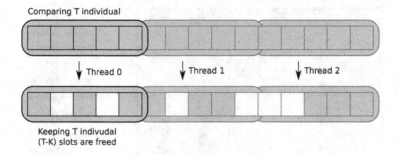

Fig. 1. Parallel individual comparison by dispar operator

single thread. The K best individuals are kept, the $(T - K)$ other are deleted. The released slots are reused to store the children for the next generation. The tournaments can be performed by $(\mu + \lambda)/T$ threads in parallel, where λ (resp. μ) is the number of parents (resp. children).

Using this operator allows to fully parallelize any evolutionary algorithm (by simulating a $(\mu + \lambda)$-ES) without the need to execute a serial reduction of the population on the CPU.

3   CUDA Hardware

A GPGPU is a processor containing a large number of cores (typically several hundreds). These computing units are grouped into Single Instruction Multiple Data (SIMD) core bundles, which execute the same instruction at the same time, applied to different data. In CUDA, these units are called Multi-Processors (MP).

Recent Fermi processors have 32 cores that execute each instruction only once. Thus, an instruction is executed by 32 different threads (called a warp), on possibly different data.

NVIDIA hardware implements synchronization mechanisms between the cores of the same warp but there is no explicit mechanism to synchronize different warps. They are implicitly synchronized at the end of the execution of a GPGPU program.

The GPGPU card contains a large global memory that can be accessed by the CPU thanks to Direct Memory Access (DMA) transfers. In addition, each MP embeds a few kilobytes of memory (16 or 48KB, depending on the card), that are only accessible by the MP cores. This extremely fast memory is called shared memory, and replaces what should be a cache to access the global memory. It is implemented using 16 banks that should not be accessed by multiple threads at the same time. The shared memory has a switch which handles complex access patterns. Indeed, if some threads of the same warp attempt to access the same bank, memory accesses are serialized. Otherwise, if threads access different banks, the switch provides simultaneous access to the shared memory.

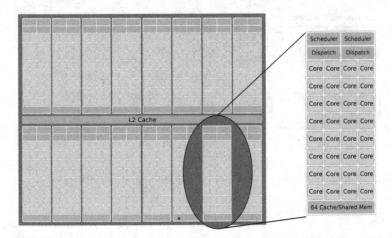

Fig. 2. Architectural details of a Fermi GPGPU Card. GPGPU chip and detail of a Multi-processor.

Without cache, a memory access is very slow (in the order of several hundreds of cycles) and due to its limited size, shared memory can not always compensate for the lack of cache. A second mechanism complementing the shared memory is a hardware thread scheduling mechanism. It is comparable to the Intel Hyper-Threading mechanism, that can be found on Pentium IV or Core i7, but at a much higher level. These units switch frozen warps, for example on memory reads, with other warps, which are ready for execution. Scheduling units can choose among a set of w warps ($w = 32$), which can come from different blocks of threads per MP. The blocks are groups of threads that are alloted to one MP, e.g. the minimal pool of threads to schedule. The size of this scheduling pool can explain why GPGPUs are not efficient with only a few threads, as shown in [6].

The structure of the cores and the availability of hardware scheduling units give GPGPUs a SPMD (Single Program Multiple Data) / SIMD structure, as a warp among w runs in parallel on each multi-processor. Indeed, the threads in a warp must necessarily execute the same instruction, because they are executed by the cores of the same MP at the same time. However, each warp can execute a different instruction without causing any divergence among the cores, because only one warp is executed at a time. In addition, the MPs can execute different instructions on their respective cores at the same time.

Finally, the bus to global memory is particularly large (for a 480GTX it is 384bits). The correct use of the memory bus allows to obtain a high theoretical bandwidth (177GB/s for a 480GTX) if, when accessing the global memory, the requested data is loaded, as well as neighbouring data, up to the width of the bus. If memory accesses of a warp are aligned (neigbouring threads reading neighbouring data), the number of memory accesses is reduced compared to totally random schemes and bandwidth is better utilised.

4   EAs on GPGPUs

Parallelization of evolutionary algorithms can tackle problems with larger search spaces than those accessible to sequential algorithms because they can deal with much larger population sizes for no extra temporal cost. The EASEA approach (evolutionary engine on CPU, parallel evaluation on remote processors) allows to benefit from GPGPUs on problems with heavy evaluation functions [6,5].

This gives two advantages: first, a heavy evaluation function compensates for the transfer of the individuals to the computation cores (GPGPU or remote nodes). Secondly, a CPU intensive evaluation makes the parallelization of the evolutionary engine less critical. Indeed, if the evaluation function takes 99% of the total time, it is possible to obtain an acceleration of up to 100x, by parallelizing the evaluation function only, according to Amdahl's law.

But these principles restrict parallelization to algorithms with heavy evaluation functions. Indeed, if the evaluation step takes 50% of the total execution time, even if parallelization makes it virtually instantaneous, the global speedup will only be x2.

In this paper, two approaches are presented that run the whole evolutionary algorithm on the GPGPU architecture. The first uses the DISPAR population reduction operator. The second focuses on improving memory usage, but is restricted to the generational algorithm, therefore eliminating the reduction phase (parents are simply replaced by children).

4.1   DISPAR GPU Evolutionary Algorithm

Population Organization. As explained in 2.2, DISPAR requires that the individuals are randomly distributed in the population, *i.e.* there must not exist a clear correlation between spatially related individuals. This is done here, by keeping a global population mixing parents and children.

This population is stored in the global memory of the card. Accessing the individuals cause memory latencies but this memory is less constrained in terms of space than the shared-memory.

A thread is dedicated to an individual. A population organization as presented in the top of Figure 3 allows to benefit from cache memory on standard single-core processors. On SIMD processors, the cores access the same gene of different individuals at the same time. A population organization as the one presented on the bottom of fig. 3 should allow to benefit from the large memory bus. For instance, accessing all genes 0 on different individuals takes fewer loading operations.

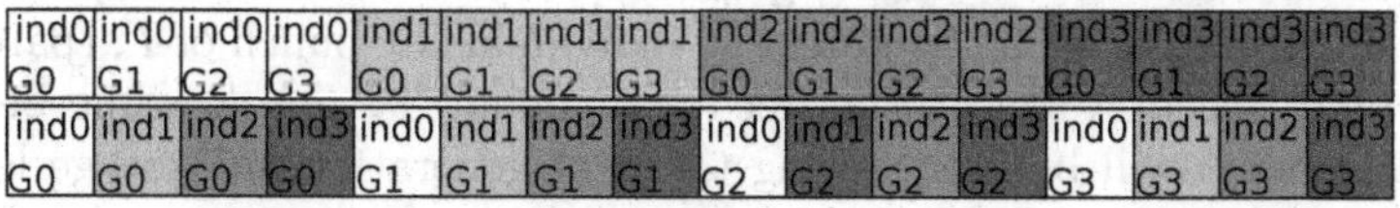

ind0	ind0	ind0	ind0	ind1	ind1	ind1	ind1	ind2	ind2	ind2	ind2	ind3	ind3	ind3	ind3
G0	G1	G2	G3	G0	G1	G2	G3	G0	G1	G2	G3	G0	G1	G2	G3
ind0	ind1	ind2	ind3	ind0	ind1	ind2	ind3	ind0	ind1	ind2	ind3	ind0	ind1	ind2	ind3
G0	G0	G0	G0	G1	G1	G1	G1	G2	G2	G2	G2	G3	G3	G3	G3

Fig. 3. Classical population layout *vs.* population organization by gene

Initialization. The algorithm initialisation phase requires the participation of both CPU and GPGPU. The CPU allocates buffer dedicated to storing individuals on the GPU card and launches the GPGPU procedures. Buffers are needed to store the genomes, the mutation coefficients, the list of free slots and the random generators data, as explained below.

The population is initialized using the parallel random number generator library now provided by CUDA.

In order to use the DISPAR reduction operator, it is necessary to maintain a list of free slots for future children. These locations are uniformly distributed throughout the population. These free locations are referenced in a dedicated table. This table contains, for each thread, a slot where the thread will place the child it will create.

After the replacement step it is still necessary to synchronize the cores, to be sure that all children have been produced. This global synchronisation is done by returning to the CPU at every generation.

Evolving Step. Each thread executes a GPGPU evolutionary procedure. This procedure performs the creation of a child, into the free slot allocated to the current thread and then executes DISPAR to determine a number of free slots for a given number of threads.

Two tournament operators are run in order to choose the two parents. These operators are classical n-ary tournament operators. As the operator randomly choses the individuals to compare, the cores execute the same instruction at the same time, but on non-contiguous data.

The crossover accesses all the genes of both parents and computes a weighted centroid (barycentric crossover) for each of them. This means that there is no divergence between threads of a single warp. But accesses to the genes are very likely to be non-contiguous in memory, because parents are coming from the tournament operator.

The mutation operator takes as input a location and applies a Schwefel self-adaptive mutation [12]. A tosscoin function tells whether a gene should be mutated or not. This causes divergences between the cores of a warp, if some threads do not mutate a gene while the other threads do. However, this divergence creates two execution paths, one of which being empty.

Reduction. As explained in section 2.2, the reduction step requires the population to be distributed randomly, which can be maintained by randomly choosing the parents of an individual within the whole population, the resulting child being placed in this same population. The population is initialized from a size $(\mu + \lambda)$. μ slots are initialized while λ are left free, alternately. These free slots will be used to store the children, ensuring their equal distribution in the population.

The reduction algorithm compares the individuals by bunch of T, (parents or children) and selects λ slots to be freed, *i.e.* among T, $T - K$ while be freed. Again, these slots will be used during the next generation, as storage location for the new children.

This way, there is no clear correlation (sibling or parentage) between spatially related individuals, which induces a memory bandwidth waste because neighbor threads handle non-neighbor individuals.

4.2 Generational Algorithm

The second algorithm reuses all the phases of the first, except the reduction phase, that is no longer needed in the case of a generational algorithm. But thanks to predictable memory accesses, the behaviour of some of these functions changes and the computing performances can be improved.

Population Organization. The DISPAR-based algorithm exhibits a lack of memory alignment. Indeed, its reduction procedure requires the free slots to be distributed randomly in the population. In the case of a generational algorithm, this constraint does not need to be respected, as the new population will totally replace the previous one.

It is therefore possible to place individuals in an appropriate manner in the memory. To do this, two different populations are used. In a generation, a buffer is used as the parent population, while the other will receive the created children. At the end of the generation, when returning to the CPU, the buffers are exchanged.

In both buffers, the genes of the individuals are placed in the memory using the same technique as for the previous algorithm and described on the bottom of fig. 3.

Evolutionary Engine. Each generation ends before the replacement phase. The parent selection operator uses the table of fitness. It randomly picks individuals in the population and compares their fitness. Here, the alignment of individuals in memory has no interest.

The crossover operator uses the individuals chosen by the tournament operator. As they can be anywhere in memory, the alignment has no positive effect on parents reading. However, neighbouring threads (for example, those of a warp) will produce children who are neighbours in the second buffer. The alignment accelerates this step.

The mutation function works on children produced by the crossover function. Therefore, neighbour threads will work on neighbour individuals.

Finally, the evaluation function uses children. As for mutation, individuals are accessed in an aligned manner from the memory point of view.

To summarize, parents selection and crossover operators work on non-neighbour individuals and therefore use the memory in a non-aligned way. Operators working on children access memory in an aligned manner.

5   Experiments

To illustrate the behaviour of the algorithms presented above, various tests are performed on a set of artificial problems, on an Intel$^{(R)}$ Core$^{(TM)}$ i7 950 CPU clocked at 3.07GHz running under Linux 2.6.32, 10.04.2 ubuntu with NVIDIA driver 260.19.26. The GPGPU card is a NVIDIA GTX480.

Comparisons are between a parallel algorithm on the GPU card and a sequential algorithm on the CPU, therefore, only one core is used on the CPU.

5.1 Benchmarks

Two well-known toy problems are used, to show the achievable speedup. The first problem is the optimization of the Rosenbrock function. It was used because it is very short to compute and therefore takes longer to execute when the evaluation step only is parallelized on a GPU card, because of the general overhead induced by the paralellization (population transfer to the GPU card, . . .). The dimension of the problem varies between 10 and 80. Equation 1 presents the Rosenbrock function used here.

$$R(x) = \sum_{i=1}^{dim}[(1 - x_i)^2 + 100(x_{x+1} - x_i^2)^2]\tag{1}$$

The second problem is to optimize the multi-dimensional Weierstrass function, which in theory is an infinite sum of sines. In practice, the number of sums allows to vary the computational load of the evaluation function. In the following tests, the number of sum iterations is varies between 10 and 80. Here, the number of dimensions is fixed to 40. This problem was also used in [6,5]. Finally, Equation 2 presents the variant that has been used in the next experiments.

$$W_{b,h}(x) = \sum_{i=1}^{dim}\sum_{j=1}^{iter} b^{-jh}sin(b^j x_i) \text{ with } b > 1 \text{ and } 0 < h < 1\tag{2}$$

Achieved Speedups. The following curves compare executions on CPU (ezCPU), with parallel evaluations on GPU (ezGPU), with DISPAR-GPU (DISPAR-GPU) and generational GPU (genGPU) algorithms. Both ezGPU and ezCPU algorithms are implemented using the same EASEA code.

Rosenbrock is a typical problem where parallelizing the evaluation function only on the GPU does not bring any improvement, as seen on fig. 4(a). With DISPAR-GPU, speedups starts between 6 and 16x (cf. fig. 4(b)) and reach 200x for a small-sized problem, with a relatively large population (10 dimensions and ≈ 245000 individuals). Above 10 dimensions the speedups fall, probably due to competition problems for the GPGPU memory cache.

For genGPU, the speedup is lower for a small population and small-sized individuals as shown in fig. 4(c), but rapidly increases up to 265× on 10 dimensions with ≈ 245000 individuals.

Speedup curves are different on the more computation intensive Weierstrass problem (cf. fig. 5(a) and 5(b)). The speedup is more interesting for a small

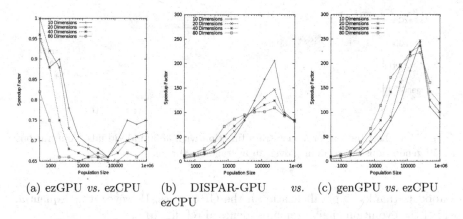

(a) ezGPU *vs.* ezCPU   (b) DISPAR-GPU   *vs.*   (c) genGPU *vs.* ezCPU
                                    ezCPU

Fig. 4. Speedup factor for 10 to 80 dimensions Rosenbrock problem

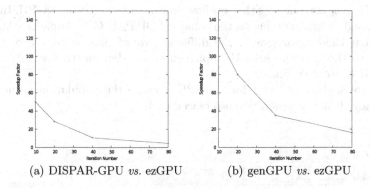

(a) DISPAR-GPU *vs.* ezGPU   (b) genGPU *vs.* ezGPU

Fig. 5. Speedup factor for Weierstrass *w.r.t* the number of iterations for a 40 dimensions problem, with a population 32768 individuals

number of iterations. However, when this number increases, the evolutionary engine becomes proportionally less important in the overall execution time. The individuals evaluation speedup should be substantially the same between ezGpu, DISPAR-Gpu and genGpu, only the evolutionary engine can be accelerated for the two last. The speedup is therefore higher at the beginning and tends to decrease, as long as the evolutionary engine loses its significance in execution time.

Study of Time Distribution. The comparison between ezCPU (sequential execution on one CPU core implemented in EASEA) and ezGPU (EASEA parallelization of the evaluation step on GPU card as described in [6]) shows that evaluation time (60% of the total time in the ezCPU implementation) virtually

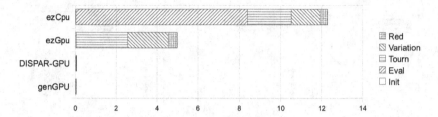

Fig. 6. Time distribution for Weierstrass function with 4800 individuals, 10 iterations and 50 generations (in seconds), over an average of 20 runs

disappears thanks to parallelizatin on the GPU cores. However, the sequential part of the evolution engine remains identical (cf. fig. 6).

When DISPAR-GPU and genGPU are used, both evaluation and evolution engine execution times are collapsed by parallelization.

Figure 7 shows these same measurements, as a percentage of the total execution time. Here again, one sees that evaluation time disappears on ezGPU. Interestingly enough, when zooming on the collapsed DISPAR-GPU implementation, one sees that the proportions of the different evolutionary steps are roughly equivalent to those of the ezCPU implementation, showing that ezCPU and DISPAR-GPU are very similar.

Things are slightly different for genGPU because the population reduction step is removed (it is a generational replacement).

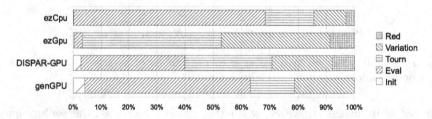

Fig. 7. Same as above, but in percentage of execution time

Quality of the Algorithms (Fitness Evolution) ezGPU, DISPAR-GPU and genGPU are different algorithms, therefore, one can expect them to solve the problem with a different efficiency. Fig. 8 shows the evolution of the best individual among a population of 30720 individuals on the Weierstrass benchmark on 10 dimensions with 10 iterations, over 20 runs. ezGpu and DISPAR-Gpu roughly follow the same trend because their population reduction operator (tournament and DISPAR tournament) are similar. genGPU is not elitist (hence the increase in best fitness in the first generations) and the lack of convergence towards 0 as the children are not always better as their parents.

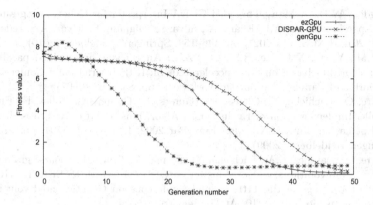

Fig. 8. Evolution of the best individual for the three GPGPU algorithms

6 Conclusion and Future Works

Two fully parallel evolutionary algorithms have been presented, that were implemented on a GPGPU card. They allow to remove the sequential part of the evolutionary algorithm that was the bottleneck of the EASEA parallelization of evolutionary algorithms proposed in [6,5]. Parallelization on GPU cards becomes interesting even on very lightweight evaluation functions such as Rosenbrock, with speedups up to about 200, with a parallel DISPAR tournament, or up to about 250 on a generational replacement, with however its disadvantages (no elitism).

The lack of a global synchronization mechanism on GPGPU, still imposes to go back to the CPU at every generation for this only purpose (the cores or the CPU are mostly left idling during the algorithm).

These speedups have been obtained with algorithms that are as standard as possible, allowing to compare their results with purely CPU implementations (as shown in fig. 8). Their behaviour is slightly different from standard ones, that can still largely benefit from GPU parallelization if the evaluation step is heavy to compute. However, more experiments have to be conducted on how our algorithm compares with standard $(\mu + \lambda)$-ES.

For lightweight evaluation function, these adaptations of standard algorithms may be interesting for real-time algorithms where low consumption embedded GPGPU cards can be used for a very high computation/consumption ratio. Their integration into the EASEA platform is in progress.

References

1. Amdahl, G.: Validity of the single processor approach to achieving large scale computing capabilities. In: Proceedings of the Spring Joint Computer Conference, April 18-20, pp. 483–485. ACM, New York (1967)
2. Fok, K.L., Wong, T.T., Wong, M.L.: Evolutionary computing on consumer graphics hardware. IEEE Intelligent Systems 22(2), 69–78 (2007)

3. Langdon, W.B.: A Many Threaded CUDA Interpreter for Genetic Programming. In: Esparcia-Alcázar, A.I., Ekárt, A., Silva, S., Dignum, S., Uyar, A.Ş. (eds.) EuroGP 2010. LNCS, vol. 6021, pp. 146–158. Springer, Heidelberg (2010)
4. Li, J.M., Wang, X.J., He, R.S., Chi, Z.X.: An efficient fine-grained parallel genetic algorithm based on GPU-accelerated. In: IFIP International Conference on Network and Parallel Computing Workshops, pp. 855–862 (2007)
5. Maitre, O., Lachiche, N., Clauss, P., Baumes, L., Corma, A., Collet, P.: Efficient Parallel Implementation of Evolutionary Algorithms on GPGPU Cards. In: Sips, H., Epema, D., Lin, H.-X. (eds.) Euro-Par 2009. LNCS, vol. 5704, pp. 974–985. Springer, Heidelberg (2009)
6. Maitre, O., Baumes, L.A., Lachiche, N., Corma, A., Collet, P.: Coarse grain parallelization of evolutionary algorithms on GPGPU cards with EASEA. In: GECCO 2009: Proceedings of the 11th Annual Conference on Genetic and Evolutionary Computation, pp. 1403–1410. ACM, New York (2009)
7. Maitre, O., Krüger, F., Querry, S., Lachiche, N., Collet, P.: EASEA: Specification and execution of evolutionary algorithms on GPGPU. Soft Computing - A Fusion of Foundations, Methodologies and Applications, Special Issue on Evolutionary Computation on General Purpose Graphics Processing Units, 1
8. Maitre, O., Querry, S., Lachiche, N., Collet, P.: EASEA parallelization of tree-based genetic programming. In: Fogel, et al. (eds.) IEEE CEC 2010, pp. 1–8. IEEE (2010)
9. Maitre, O., Sharma, D., Lachiche, N., Collet, P.: DISPAR-Tournament: A Parallel Population Reduction Operator That Behaves Like a Tournament. In: Di Chio, C., Cagnoni, S., Cotta, C., Ebner, M., Ekárt, A., Esparcia-Alcázar, A.I., Merelo, J.J., Neri, F., Preuss, M., Richter, H., Togelius, J., Yannakakis, G.N. (eds.) EvoApplications 2011, Part I. LNCS, vol. 6624, pp. 284–293. Springer, Heidelberg (2011)
10. Pospichal, P., Jaros, J., Schwarz, J.: Parallel Genetic Algorithm on the CUDA Architecture. In: Di Chio, C., Cagnoni, S., Cotta, C., Ebner, M., Ekárt, A., Esparcia-Alcazar, A.I., Goh, C.-K., Merelo, J.J., Neri, F., Preuß, M., Togelius, J., Yannakakis, G.N. (eds.) EvoApplicatons 2010, Part I. LNCS, vol. 6024, pp. 442–451. Springer, Heidelberg (2010)
11. Robilliard, D., Marion-Poty, V., Fonlupt, C.: Genetic programming on graphics processing units. Genetic Programming and Evolvable Machines 10(4), 447–471 (2009)
12. Schwefel, H.P.: Numerical Optimization of Computer Models. Wiley, Chichester (1981)
13. Yu, Q., Chen, C., Pan, Z.: Parallel Genetic Algorithms on Programmable Graphics Hardware. In: Wang, L., Chen, K., S. Ong, Y. (eds.) ICNC 2005, Part III. LNCS, vol. 3612, pp. 1051–1059. Springer, Heidelberg (2005)

A Multilevel Tabu Search with Backtracking for Exploring Weak Schur Numbers

Denis Robilliard, Cyril Fonlupt, Virginie Marion-Poty, and Amine Boumaza

Univ Lille Nord de France,
ULCO, LISIC, BP 719,
62228 CALAIS Cedex, France
{robilliard,fonlupt,poty,boumaza}@lisic.univ-littoral.fr

Abstract. In the field of Ramsey theory, the *weak Schur number* $WS(k)$ is the largest integer n for which their exists a partition into k subsets of the integers $[1, n]$ such that there is no $x < y < z$ all in the same subset with $x + y = z$. Although studied since 1941, only the weak Schur numbers $WS(1)$ through $WS(4)$ are precisely known, for $k \geq 5$ the $WS(k)$ are only bracketed within rather loose bounds. We tackle this problem with a tabu search scheme, enhanced by a multilevel and backtracking mechanism. While heuristic approaches cannot definitely settle the value of weak Schur numbers, they can improve the lower bounds by finding suitable partitions, which in turn can provide ideas on the structure of the problem. In particular we exhibit a suitable 6-partition of $[1, 574]$ obtained by tabu search, improving on the current best lower bound for $WS(6)$.

Keywords: tabu search, weak Schur numbers, optimization.

1 Introduction

1.1 Mathematical Description

Ramsey theory is a branch of mathematics that studies the existence of orderly sub-structures in large chaotic structures: *"complete disorder is impossible"*[1]. Many questions in Ramsey theory remain open, and they are typically phrased as *"how many elements of some structure must there be to guarantee that a particular property will hold?"*.

In this paper we focus on the following Ramsey theory problem: finding weak Schur numbers. A set P of integers is called *sum-free* if it contains no elements $x, y, z \in P$ such that $x + y = z$. A theorem of Schur [1] states that, given $k \geq 1$, there is a largest integer n for which the integer interval set $[1, n]$ (i.e. $\{1, 2, \ldots, n\}$) admits a partition into k sum-free sets. This largest integer n is called the k-th Schur number, denoted by $S(k)$.

Adding the assumption that x, y, z must be *pairwise distinct*, i.e. $x \neq y, y \neq z, x \neq z$, we define *weakly sum-free* sets, and a similar result was shown by

---

[1] Quote attributed to Theodore Motzkin.

J.-K. Hao et al. (Eds.): EA 2011, LNCS 7401, pp. 109–119, 2012.

Rado [2] : given $k \geq 1$, there is a largest integer n for which the interval set $[1, n]$ admits a partition into k weakly sum-free sets: this largest integer n is the weak Schur number denoted $WS(k)$.

For example, in the case $k = 2$, a weakly sum-free partition of the first 8 integers is provided by:

$$\{1, 2, 3, 4, 5, 6, 7, 8\} = \{1, 2, 4, 8\} \cup \{3, 5, 6, 7\}$$

It is straightforward to verify that increasing the interval to $[1, 9]$ yields a set that does not admit any weakly sum-free partition into 2 sets, thus we have $WS(2) = 8$.

Very few exact values are known for $WS(k)$ (or for $S(k)$):

- $WS(1) = 2$ and $WS(2) = 8$ are easily verified
- $WS(3) = 23$, $WS(4) = 66$ were shown by exhaustive computer search in [3].

The same study [3] shows the following lower bound: $WS(5) \geq 189$, while a much older note by Walker [4] claimed, without proof, that $WS(5) = 196$. Recently a paper by Eliahou, Marín, Revuelta and Sanz [5] provided a weakly sum-free partition of the set $[1, 196]$ in 5 sets, confirming Walker's claim that $WS(5)$ is at least as large as 196, and also gave a weakly sum-free partition of $[1, 572]$ in 6 sets, thus establishing that:

$$WS(5) \geq 196$$

$$WS(6) \geq 572$$

For the general case $k \geq 5$, the best known results are a lower bound from Abbot and Hanson [6] and an upper bound by Bornsztein [7]:

$$c89^{k/4} \leq S(k) \leq WS(k) \leq \lfloor k!ke \rfloor$$

with c a small positive constant.

1.2   Tabu Search

Local search methods are among the simplest iterative search methods. In local search, for a given solution i a neighborhood $N(i)$ is defined, and the next solution is searched among the solutions in $N(i)$.

Tabu search (TS) is based on local search principles. It was introduced by Glover [8]. Unlike the well-known hill-climber search, the current solution may deteriorate from one iteration to the next to get out of a local optimum. In order to avoid possible cycling in the vicinity of a local optimum, tabu search introduces the notion of tabu list, which prohibits moves towards solutions which were recently explored. The duration for a move (or attribute of a move) to remain tabu is called the tabu-tenure. However, the tabu list can be overridden if some conditions, called aspiration criterion, are met. For instance, a typical aspiration criterion is the discovery of a new best solution (better than any previously visited solution): the move towards such a solution is retained even if it is forbidden by the tabu list. TS was improved with many other schemes like intensification (to focus on some part of the search space) or diversification (to explore some new part of the search space). Details about tabu search and enhancements of tabu search can be found in [9,10].

In the case of the weak Schur numbers, the search space size, including non feasible solutions, is $k^{WS(k)}$, where $WS(k)$ is at least polynomial in k. Indeed, even with small values of k a basic tabu search approach is unable to solve the problem in reasonable time limits, as shown in Section 3.

The basic idea of a multilevel approach is to create a sequence of successive coarser approximations of the problem to be solved in reverse order. The multilevel scheme was already successfully used for solving some combinatorial optimization problems [11]. In our case, we skip the explicit construction of the sequence, since it can be obtained naturally by considering each $WS(k)$ problem as a coarser version of step $WS(k + 1)$, as explained later in Section 2.1: we start with $WS(2)$ as the coarsest level and we solve the sequence of problems of increasing difficulty obtained by adding successive partitions, i.e. increasing the value of k. Our method also differs from the original multilevel in that it does not create new temporary coarse variables that encapsulate several variables of the refined problem. Here a coarser version of the problem is coarse in the sense that it contains only a small part of the variables from the complete refined version. Nonetheless we think that it borrows enough inspiration from this scheme (increasing number of variables in several large steps, earlier choices remaining fixed and interfering with later ones) to still be qualified as multilevel, and we will keep this denomination through this paper.

Anyway the multilevel scheme proved useful but not yet sufficient, due to the fact that some intermediate stage may not possibly be refined to the next level. We added a backtracking mechanism to allow the search algorithm to rewind back and introduce more diversity into previous levels.

The rest of the paper is organized as follows. In Sect. 2, we detail our approach, with a description of the multilevel framework and the backtracking mechanism. In Sect. 3, we present the new results that were discovered using our enhanced tabu scheme.

2 Detailed Framework

2.1 Problem Description and Multilevel Paradigm

As said in the Introduction, our aim is to find weakly sum-free partitions for the currently known lower bounds for $WS(5)$ and $WS(6)$, and possibly improve (increase) these lower bounds.

In [3], the authors pointed out that there exists 29931 different possible solutions to the $WS(4)$ problem (there are 29931 partitions of $[1, WS(4)]$ into four weakly sum-free subsets). Eliahou and Chappelon [12] found one interesting point: in all these 29931 solutions, the first 23 integers are always stored in only

3 of the 4 sets. Indeed, since $WS(3) = 23$, all the partitions that solve the $WS(4)$ problem are **extensions** of partitions that solve the $WS(3)$ problem. They also noticed that only 2 of the 3 possible weakly sum-free partitions of $[1, WS(3)]$ can be extended to form a weakly sum-free partition of $[1, WS(4)]$.

Eliahou *et al.* in [5] used the `march-pl` SAT solver [13], with a suitable translation of the problem in CNF form to study weak Schur numbers. As the search space increases exponentially with higher values of k for $WS(k)$, the authors could not explore in an exhaustive way all possible solutions. So they conjectured that the property that was exhibited from $WS(3)$ to $WS(4)$ could also hold true from $WS(4)$ to $WS(5)$. So when searching for $WS(5)$, Eliahou *et al.* first searched for solutions to $WS(4)$, then froze the 66 first integers into their respective sets (since $WS(4) = 66$), before searching for $WS(5)$. Conjecturing $WS(5) = 196$ leaves 130 remaining integers to assign, thus greatly reducing the size of the search space. Anyway this reduction was not yet enough to use the `march-pl` solver, and the authors of [5] had to bet on the position of part of the remaining 130 numbers in order to lower even further the computational cost. These hypotheses led them to discover a 5-partition of $[1, 196]$, proving $WS(5) \geq 196$ as part of G. W. Walker's claim. Using further placement hypotheses, they also obtained a 6-partition of $[1, 572]$, proving $WS(6) \geq 572$.

In this paper, we use the same multilevel approach for studying $WS(5)$ and $WS(6)$ by way of tabu search. When a weakly sum-free 4-partition is found for $[1, WS(4)]$, the solution is kept as an un-mutable part of the 5-partitions of $[1, x]$, with $x \geq 196$ the best known lower bound for $WS(5)$. When such a weakly sum-free 5-partition is found, it is again kept unchanged to form the basis of 6-partitions of $[1, y]$, with $y \geq 572$ the best current value for $WS(6)$.

Because some solutions at level k may be impossible to extend to give a level $(k + 1)$ solution, we enhance this technique with the following scheme: when a fixed number of trials fails to give a weakly sum-free $(k+1)$-partition, we assume that we may be trapped in a local optimum. Thus the algorithm is allowed to backtrack to the level k to find a new weakly sum-free k-partition. The details of the algorithm can be found in the next subsection.

2.2 Multilevel Tabu Search with Backtracking

The main components of our tabu search are given here, while the pseudo-code is available in Table 2.

Representation: a solution to the $WS(k)$ problem consists in a partition of the set $[1, n]$ into k sets (possibly non weakly sum-free).

Fitness function: the fitness function is simply the number of integers that are involved in a constraint violation. For example, if 2, 8 and 10 are in the same set, this implies the fitness is increased by 3 (as $2 + 8 = 10$). A perfect solution has a fitness $= 0$, i.e. whenever no constraint violations remain.

Formally, let c be a partition of $[1, n]$ into k sets (i.e. a solution), and let $P_i(c)$ be the ith set of the k-partition of c:

$$f(c) = 3 * \sum_{i=1}^{k} |\{x \in P_i(c) \mid \exists y, z \in P_i(c), x < y < z, x + y = z\}|$$

Move operator: a move operator should transform a solution in such a way that it implies only a small change in the search space. In our case, we chose to move an integer from a partition to another. This is performed in two steps: removing the integer from its origin set, and adding it to its destination set (different from its origin set).

The move operator defines the notion of move gain, indicating how much the fitness of a solution is improved (or downgraded) according to the optimization objective when moving an integer to another set. Here a negative move gain (fitness change) characterizes a move that leads to a better solution (minimization problem). The basic idea of the tabu search (ignoring tabu list and implementation details) is to select the move leading to the best solution, so the search complexity is $O(WS(k) * k)$.

A cheap computation of the move gain, if available, allows a fast, incremental evaluation of the fitness of solutions, hence a fast exploration of the neighborhood in search for the best non tabu neighbor.

In order to accelerate the computation of the move gain, a vector of constraint count is maintained for each set of the partition. This vector associates a number of constraints to every integer in the interval to be partitioned. More precisely, for each pair of integers x and y in the set, we associate one constraint to $x + y$, $y - x$ and $x - y$ (provided these positions are in the interval $[1, n]$ and $y \neq (x - y)$, $x \neq (y - x)$ to respect the pairwise distinct numbers assumption). This constraint count is performed even if the integer is not currently in the set, as illustrated in Table 1: this is akin to *forward arc checking* in constraint propagation algorithms.

- The move gain is obtained by subtracting the constraint count associated to the former position of the integer from the constraint count of its new position.
- The update of the constraint vector can be easily performed:
  - When removing an integer y from a set s, for each other integer x in set s, the number of constraints is decremented by one for all the following indices : $x + y$, $y - x$ and $x - y$ (provided these positions are in the interval $[1, n]$ and $y \neq (x - y)$, $x \neq (y - x)$ to respect the pairwise distinct numbers assumption) — these changes are done in the constraint vector associated to set s;
  - When adding an integer y to a set, the reverse operation is done, i.e. incrementing the constraint count by one at the same positions;
  - The two previous operations have a computational complexity in $O(WS(k))$.

Neighborhood: as explained before, we are exploring partitions of $[1, WS(k)]$ by extending already found weakly sum-free solutions for the $WS(k - 1)$

Table 1. Illustration of the constraint count vectors for a 2-partition of $[1,8]$

set 1	indices	1	2	3	4	5	6	7	8
	integers in subset	X	X		X		X		
	constraint count	0	1	2	1	2	1	1	1
set 2	indices	1	2	3	4	5	6	7	8
	integers in subset			X		X		X	X
	constraint count	1	2	1	1	1	0	0	1

Explanation (on set 2):
- integer 1 has 1 pending constraint since if we move 1 to set 2 a conflict will occur with 7 and 8;
- integer 2 has 2 pending constraints due to the presence of respectively $\{3,5\}$ and $\{5,7\}$ in set 2;
- integers 3, 5 and 8 have 1 actual constraint each, since they are involved in the same constraint violation (and only this one);
- integer 4 has 1 pending constraint due to the presence of the pair $\{3,7\}$ in set 2;
- integers 6 and 7 have no pending (nor actual) constraint (this notably implies that we can insert 6 in set 2 without degrading fitness).

level. So only the integers in $[WS(k-1)+1, n]$ are allowed to be moved in a $WS(k)$ level solution. The neighborhood of a solution is then the set of all possible solutions that can be reached in one application of the move operator on the allowed subset of integers.

Formally, the neighborhood $N(c)$ of a k-partition c is the set of solutions c':
$$N(c) = \{c' \mid \exists\,!\, x \in [WS(k-1)+1, n], i, j \in [1, k], i \neq j, x \in P_i(c), x \in P_j(c')\}$$

Tabu List and Tabu Tenure Management: each time an integer i is moved from a partition to another partition, it is forbidden to move again the integer i in any partition for the next tt iterations. Each time we launch the tabu search on a given sub-level of the problem, we set dynamically this tabu tenure tt to a random number between 2 and half the number of integers allowed to move.

Aspiration Criterion: the aspiration criterion overrides the tabu list when a solution is better than the currently best-known solution.

Perturbation Strategy: a random neighbor can always be chosen with a small probability, or else the algorithm chooses the non-tabu neighbor with the best gain move (taking into account the aspiration criterion). As said above we keep the previous sub-level solution unchanged in order to focus on critical elements and narrow the search space. To complement this search strategy, a rewind back mechanism is applied when no improvement to the current solution is detected for a given number of restarts of the tabu (see Sect. 3 for details). In that case, we conjecture that the algorithm is working on a non-extensible solution and we settle back to the previous $(k-1)$ sub-level problem.

Table 2. Pseudo-code for the tabu search algorithm

Step 1: Initialization
Initialize solution s to $WS(2)$
Set k = 3 {*initial number of partitions*}

Step 2: Extension of solution
Initialize *iteration_counter, tabu_list, tabu_tenure*
Initialize n to either $WS(k)$ if known, or the expected value of $WS(k)$
for each integer i such that $WS(k-1) < i \leq n$
   randomly assign i to a subset in $[1, k]$
end for
f = fitness(s)   { *evaluate the fitness of s* }
Repeat
   { *either a random move or a neighborhood exploration* }
   if (random in $[0, 1] > 0.9$)
      perform a random move
   else   { *apply neighborhood search to s* }
      Initialize δ to ∞ {*worst possible move gain*}
      for each integer i such that $WS(k-1) < i \leq n$
         { P_i is the current subset holding i }
         for all possible subset $P' \neq P_i$
            evaluate δ the fitness improvement when moving i to P'
            if ((δ is the greater so far) &&
               (move not in *tabu_list* || aspiration criterion is true))
               record this move as best_move
            end if
         end for
      end for
      perform best_ move
   end if
   update fitness f   { *quick computation using move gain* }
until stop condition
{ *either $WS(k)$ is solved or maximum # iterations is reached* }

Step 3: Extension or rewind back
if (s solution to $WS(k)$)
   k = k + 1
   Go to Step 2
else
   if (maximum number of trial runs for $WS(k)$ is not yet reached)
      runs = runs + 1
      Go to Step 2
   else
      { *seemingly this solution is not extensible, rewind back* }
      k = k - 1
      reset number of trial for $WS(k)$
      Go to Step 2
   end if
end if

3   Experimental Results

3.1   Experimental Setting

Experiments were conducted with four different types of tabu: the first one is a "classic" tabu that tried to solve directly the $WS(5)$ problem. The second scheme is the same tabu with a restart strategy. The third is a multilevel approach that first tried to solve $WS(2)$, then $WS(3)$ and so on, also using restarts. The fourth scheme is the multilevel with backtracking introduced in Sect. 2.2, referred to as "MLB tabu", starting from $WS(2)$.

In order to allow a fair comparison of the four approaches, all methods were allowed the same total of 10.000.000 fitness evaluations. For the classic tabu with restarts, the multilevel and the backtracking scheme, the maximum number of iterations for a tabu trial is set to 1.000.000 before a restart occurs. For the backtracking scheme we allow $Max\_Run = 5$ restarts before backtracking. We conducted 30 independent runs for the four approaches (each run with a maximum number of evaluations of 10.000.000).

Note that the comparison setting could be deemed slightly biased in favor of the two classic tabu versions, as they directly start with the $WS(5)$ problem unlike the other approaches which start with the $WS(2)$ problem. As a fitness evaluation for the $WS(5)$ problem is actually more time consuming than for the $WS(2)$ problem, the classic versions are in fact also allowed more computing time.

3.2   Results and Analysis

Experiments on $WS(5)$: Table 3 summarizes the results of our experiments on the $WS(5)$ problem: we call "successful run" a run that reaches the current best known bound, 196. For each scheme we report the results obtained within 30 independent runs.

Table 3. Experimental results for Schur number $WS(5)$

	Average fitness	# of successful runs
Classic tabu	17.2	0
Classic tabu with restarts	8.3	0
Multilevel tabu	6.15	1
MLB tabu	2.3	10

The running times are comparable, and even slightly in favor of the most successful methods. Indeed the first two methods work full time on $WS(5)$, i.e. on length 196 data structures, while both multilevel algorithms spend some time solving $WS(4)$, thus working on smaller data.

The multilevel backtracking mechanism outperforms the other methods considered here by a considerable margin. First, both classic versions were unable to find the current best known bound for the $WS(5)$ problem. Anyhow the restart

Table 4. Sum free 6-partition of the 574 first integers

$E_1 =$ [1, 2, 4, 8, 11, 22, 25, 31, 40, 50, 60, 63, 69, 84,
97, 135, 140, 145, 150, 155, 164, 169, 178, 183, 193, 199,
225, 258, 273, 330, 353, 356, 395, 400, 410, 415, 438, 447,
461, 504, 519, 533, 547, 556, 561, 568, 571]

$E_2 =$ [3, 5, 6, 7, 19, 21, 23, 35, 36, 51, 52, 53, 64, 65,
66, 80, 93, 109, 122, 124, 137, 138, 139, 151, 152, 153, 165,
180, 181, 182, 194, 195, 196, 210, 212, 226, 241, 251, 255,
298, 310, 314, 325, 340, 341, 354, 355, 369, 371, 384, 397,
398, 399, 411, 412, 413, 426, 440, 441, 442, 458, 472, 473,
482, 486, 498, 500, 502, 514, 515, 529, 530, 531, 542, 543,
558, 559, 560, 572, 573, 574]

$E_3 =$ [9, 10, 12, 13, 14, 15, 16, 17, 18, 20, 54, 55,
56, 57, 58, 59, 61, 62, 101, 103, 104, 107, 141, 142, 143,
144, 146, 147, 148, 149, 184, 185, 186, 187, 188, 189, 190,
191, 192, 227, 229, 230, 232, 233, 234, 235, 269, 270, 276,
317, 319, 320, 321, 322, 359, 360, 361, 363, 364, 365, 401,
402, 403, 404, 405, 406, 407, 408, 409, 443, 444, 445, 446,
448, 449, 450, 451, 476, 477, 478, 479, 483, 484, 520, 521,
522, 523, 524, 525, 526, 527, 528, 562, 563, 564, 565, 566,
567, 569, 570]

$E_4 =$ [24, 26, 27, 28, 29, 30, 32, 33, 34, 37, 38, 39,
41, 42, 43, 44, 45, 46, 47, 48, 49, 154, 156, 157, 158, 159,
160, 161, 162, 163, 166, 167, 168, 170, 171, 172, 173, 174,
175, 176, 177, 179, 284, 286, 288, 289, 292, 294, 295, 303,
304, 305, 309, 414, 416, 417, 418, 419, 420, 421, 422, 423,
424, 425, 427, 428, 429, 430, 431, 432, 433, 434, 435, 436,
437, 439, 532, 534, 535, 536, 537, 538, 539, 540, 541, 544,
545, 546, 548, 549, 550, 551, 552, 553, 554, 555, 557]

$E_5 =$ [67, 68, 70, 71, 72, 73, 74, 75, 76, 77, 78, 79,
81, 82, 83, 85, 86, 87, 88, 89, 90, 91, 92, 94, 95, 96, 98,
99, 100, 102, 105, 106, 108, 110, 111, 112, 113, 114, 115,
116, 117, 118, 119, 120, 121, 123, 125, 126, 127, 128, 129,
130, 131, 132, 133, 134, 136, 271, 274, 275, 279, 280, 282,
283, 287, 291, 296, 299, 302, 307, 308, 312, 313, 315, 452,
453, 454, 455, 456, 457, 459, 460, 462, 463, 464, 465, 466,
467, 468, 469, 470, 471, 474, 475, 480, 481, 485, 487, 488,
489, 490, 491, 492, 493, 494, 495, 496, 497, 499, 501, 503,
505, 506, 507, 508, 509, 510, 511, 512, 513, 516, 517, 518]

$E_6 =$ [197, 198, 200, 201, 202, 203, 204, 205, 206, 207,
208, 209, 211, 213, 214, 215, 216, 217, 218, 219, 220, 221,
222, 223, 224, 228, 231, 236, 237, 238, 239, 240, 242, 243,
244, 245, 246, 247, 248, 249, 250, 252, 253, 254, 256, 257,
259, 260, 261, 262, 263, 264, 265, 266, 267, 268, 272, 277,
278, 281, 285, 290, 293, 297, 300, 301, 306, 311, 316, 318,
323, 324, 326, 327, 328, 329, 331, 332, 333, 334, 335, 336,
337, 338, 339, 342, 343, 344, 345, 346, 347, 348, 349, 350,
351, 352, 357, 358, 362, 366, 367, 368, 370, 372, 373, 374,
375, 376, 377, 378, 379, 380, 381, 382, 383, 385, 386, 387,
388, 389, 390, 391, 392, 393, 394, 396]

version is better, certainly benefiting from a necessary diversification. The third algorithm, multilevel only, performs slightly better as it is able to find, once, the bound to the $WS(5)$ problem and gets an average error of 6.15 (2 triplet constraint violations remain on average). The multilevel coupled with backtracking outperforms its competitors, rediscovering the best known bound 10 times out of 30 runs, and getting the lowest average error. We recall that it is known from [5] that only 2 of the 3 partitions for $WS(3)$ can be extended to construct a successful partition for $WS(4)$. This was the motivation for introducing the backtracking mechanism, and from these results it also seems that not all $WS(4)$ partitions can be extended to $WS(5)$, explaining the difference in success ratio between the two last algorithms of Table 3.

As none of the algorithms were able to improve the 196 lower bound for $WS(5)$, this gives some confidence that it may be the exact bound.

Experiments on $WS(6)$: We also used the multilevel and backtracking version to work on the $WS(6)$ problem, obtaining several weakly sum-free 6-partitions of the set $[1, 572]$, the best known bound at the time. This algorithm was also able to find a weakly sum-free 6-partition of $[1, 574]$, setting a new lower bound to the 6th weak Schur number problem:

$$WS(6) \geq 574$$

Table 4 presents this partition for the $WS(6)$ problem.

4   Conclusions

This work shows that local search optimization techniques such as tabu search can be used to tackle hard Ramsey theory problems, allowing to refine bounds and to obtain instances of solutions that would be difficult, if not impossible, to obtain by other methods, such as exhaustive constraint propagation.

While heuristic approaches cannot definitely settle the value of weak Schur numbers, the solutions we obtained can suggest ideas on the structure of the problem. This study also gives some confidence that 196 is a very probable tight bound for $WS(5)$, while it also sets a new lower bound for $WS(6) \geq 574$.

In our case, the tabu search had to be enhanced by multilevel search and backtracking in order to successfully tackle this problem. Hybridizing our tabu search with some constraint propagation mechanisms could probably enhance further the power of this heuristic.

Tuning the tabu parameters by using a tool based on racing algorithms [14], has been initiated, although the computing time cost is large. First results on $WS(3)$ enhance the importance of allowing a ratio of random moves in the search process, and plead for a slightly larger minimum tabu tenure.

Note added in proof. The lower bound on $WS(6)$ has been recently improved by our colleagues from [5] and is now given by $WS(6) \geq 575$. This result was obtained by using a constraint solver and information drawn from our proposal partition in Table 4.

References

1. Schur, I.: Über die kongruenz $x^m + y^m \equiv z^m$ (mod p). Jahresbericht der Deutschen Mathematiker Vereinigung 25, 114–117 (1916)
2. Rado, R.: Some solved and unsolved problems in the theory of numbers. Math. Gaz. 25, 72–77 (1941)
3. Blanchard, P.F., Harary, F., Reis, R.: Partitions into sum-free sets. In: Integers 6, vol. A7 (2006)
4. Walker, G.W.: A problem in partitioning. Amer. Math. Monthly 59, 253 (1952)
5. Eliahou, S., Marín, J.M., Revuelta, M.P., Sanz, M.I.: Weak Schur numbers and the search for G. W. Walker's lost partitions. Computer and Mathematics with Applications 63, 175–182 (2012)
6. Abbott, H.L., Hanson, D.: A problem of Schur and its generalizations. Acta Arith. 20, 175–187 (1972)
7. Bornsztein, P.: On an extension of a theorem of Schur. Acta Arith. 101, 395–399 (2002)
8. Glover, F.: Tabu search - part I. ORSA Journal of Computing 1(3), 190–206 (1989)
9. Glover, F., Laguna, M.: Tabu Search. Kluwer Academic Publishers, Norwell (1997)
10. Hamiez, J.P., Hao, J.K., Glover, F.W.: A study of tabu search for coloring random 3-colorable graphs around the phase transition. Int. Journal of Applied Metaheuristic Computing 1(4), 1–24 (2010)
11. Walshaw, C.: Multilevel Refinement for Combinatorial Optimisation Problems. Annals Oper. Res. 131, 325–372 (2004)
12. Chappelon, J., Eliahou, S.: Personal communication (2010)
13. Heule, M., Maaren, H.V.: Improved version of march ks, http://www.st.ewi.tudelft.nl/sat/Sources/stable/march_pl (last accessed on May 2010)
14. López-Ibáñez, M., Dubois-Lacoste, J., Stützle, T., Birattari, M.: The irace package, iterated race for automatic algorithm configuration. Technical Report TR/IRIDIA/2011-004, IRIDIA, Université Libre de Bruxelles, Belgium (2011)

An Improved Memetic Algorithm
for the Antibandwidth Problem⋆

Eduardo Rodriguez-Tello and Luis Carlos Betancourt

CINVESTAV-Tamaulipas, Information Technology Laboratory.
Km. 5.5 Carretera Victoria-Soto La Marina, 87130 Victoria Tamps., Mexico
{ertello,lbetancourt}@tamps.cinvestav.mx

Abstract. This paper presents an Improved Memetic Algorithm (IMA) designed to compute near-optimal solutions for the antibandwidth problem. It incorporates two distinguishing features: an efficient heuristic to generate a good quality initial population and a local search operator based on a Stochastic Hill Climbing algorithm. The most suitable combination of parameter values for IMA is determined by employing a tunning methodology based on Combinatorial Interaction Testing. The performance of the fine-tunned IMA algorithm is investigated through extensive experimentation over well known benchmarks and compared with an existing state-of-the-art Memetic Algorithm, showing that IMA consistently improves the previous best-known results.

Keywords: Memetic Algorithms, Antibandwidth Problem, Combinatorial Interaction Testing, Parameter Tunning.

1 Introduction

The *antibandwidth* problem was originally introduced as the *separation number* problem by Leung et al. in connection with the multiprocessor scheduling problem [1]. Later, it has also received the name of *dual bandwidth* [2]. This combinatorial optimization problem consists in finding a labeling for the vertices of a graph $G(V, E)$, using distinct integers $1, 2, \ldots, |V|$, so that the minimum absolute difference between labels of adjacent vertices is maximized.

There exist practical applications of the antibandwidth problem which arise in various fields. Some examples are: radio frequency assignment problem [3], channels assignment and T-coloring problems [4], obnoxious facility location problem [5] and obnoxious center problem [6, 2].

The antibandwidth problem can be formally stated as follows. Let $G(V, E)$ be a finite undirected graph, where V ($|V| = n$) defines the set of vertices and $E \subseteq V \times V = \{\{i, j\} : i, j \in V\}$ is the set of edges. Given a bijective labeling

---

⋆ This research work was partially funded by the following projects: CONACyT 99276, Algoritmos para la Canonización de Covering Arrays; 51623 Fondo Mixto CONACyT y Gobierno del Estado de Tamaulipas.

J.-K. Hao et al. (Eds.): EA 2011, LNCS 7401, pp. 121–132, 2012.

function for the vertices of G, $\varphi : V \to \{1, 2, \ldots, n\}$, the antibandwidth for G with respect to the labeling φ is defined as:

$$\mathrm{AB}_\varphi(G) = \min\{|\varphi(i) - \varphi(j)| : (i, j) \in E\} \tag{1}$$

Then the antibandwidth problem consists in finding a labeling (solution) φ^* for which $\mathrm{AB}_{\varphi^*}(G)$ is maximized, i.e.,

$$\mathrm{AB}_{\varphi^*}(G) = \max\{\mathrm{AB}_\varphi(G) : \varphi \in \mathscr{L}\} \tag{2}$$

where $\mathscr{L}$ is the set of all possible labeling functions.

Leung et al. have shown that finding the maximum antibandwidth of a graph is NP-hard for general graphs [1]. Therefore, there is a need for heuristics to address this problem in reasonable time since it is unlikely that exact algorithms running in polynomial time exist for solving it in the general case.

This paper aims at developing an Improved Memetic Algorithm (IMA) for finding near-optimal solutions for the antibandwidth problem. To achieve this, the proposed IMA algorithm incorporates two distinguishing features: a fast heuristic to create a good quality initial population and a local search operator based on a Stochastic Hill Climbing algorithm. Through the use of a tunning methodology, based on Combinatorial Interaction Testing [7], the combination of both components and parameter values for IMA was determined to achieve the best trade-off between solution quality and computational effort. The performance of IMA is assessed with a test-suite, composed by 30 benchmark instances taken from the literature. The computational results are reported and compared with previously published ones, showing that our algorithm is able to consistently improve the previous best-known solutions for the selected benchmark instances.

The rest of this paper is organized as follows. In Sect. 2, a brief review is given to present some representative solution procedures for the antibandwidth problem. Then, the components of our Improved Memetic Algorithm are discussed in detail in Sect. 3. Two computational experiments are presented in Sect. 4. The first one is dedicated to determine the best parameter settings for IMA, while the second carries out a performance comparison of IMA with respect to an existing state-of-the-art Memetic Algorithm. Finally, the last section summarizes the main contributions of this work and presents some possible directions for future research.

2   Relevant Existing Procedures

Because of the theoretical and practical importance of the antibandwidth problem, much research has been carried out in developing effective heuristics for it. Most of the previous published work on the antibandwidth problem was devoted to the theoretical study of its properties for finding optimal solutions for specific cases. Polynomial time exact algorithms are known for solving some special instances of the antibandwidth problem: paths, cycles, special trees, complete and complete bipartite graphs, meshes, and tori [8, 2, 9–12].

The work of Bansal and Srivastava [13] in an exception. They proposed a Memetic Algorithm, called MAAMP, which starts by constructing an initial population through the use of a label assignment heuristic, called LAH. It builds a level structure of a graph using a random breadth first search. Then, it randomly choses to start the labeling process either by the even or the odd levels of the structure. The vertices of the graph belonging to the same level are labeled one at a time in a greedy manner. MAAMP continues performing a series of cycles called generations. At each generation, selection for mating is done by applying a binary tournament operator in such a way that each individual in the parents population participates in exactly two tournaments. Children are generated using a unary reproduction operator that constructs a level structure of a graph by employing an intermediate breadth first search to produce a new labeling. Then, mutation based on swapping two randomly selected labels is performed over the solutions in the children population. Finally, the population is updated when each child competes with its respective parent and the best of them becomes the parent for the next generation. This process repeats until the Memetic Algorithm ceases to make progress, i.e., when a better solution is not produced in a predefined number of successive generations. The authors argued that MAAMP was able to obtain optimal solutions for standard graphs like paths, cycles, d-dimensional meshes, tori, hypercubes and complement of power graphs. However, they recognize that their algorithm did not reach the optimal antibandwidth in the case of unbalanced trees and complete binary trees. Furthermore, Bansal and Srivastava only present detailed results of their experiments for a set of 30 random connected graphs.

3   An Improved Memetic Algorithm

In this section we present the implementation details of the key components of an Improved Memetic Algorithm (IMA) for solving the antibandwidth problem. For some of these components different possibilities were analyzed (see Sect. 4.2) in order to find the combination which offers the best quality solutions at a reasonable computational effort.

3.1   Search Space, Representation and Fitness Function

Given a graph $G = (V, E)$ with vertex set V ($|V| = n$) and edge set E. The search space $\mathscr{L}$ for the antibandwidth problem is composed of all possible labelings (solutions) from V to $\{1, 2, ..., n\}$, i.e. there exist $n!/2$ possible labelings for a graph with n vertices[1].

In our IMA a labeling φ is represented as an array l of integers with length n, which is indexed by the vertices and whose i-th value $l[i]$ denotes the label assigned to the vertex i. The fitness $\mathrm{AB}_\varphi(G)$ of the labeling φ is evaluated by using (1).

---

[1] Because each one of the $n!$ labelings can be reversed to obtain the same antibandwidth.

3.2   General Procedure

Our IMA implementation starts building an initial population P, which is a set of configurations having a fixed constant size $|P|$. Then, it performs a series of cycles called generations. At each generation, assuming that $|P|$ is a multiple of four, the population is randomly partitioned into ($|P|$ mod 4) groups of individuals. Within each group, the two most fit individuals are chosen to become the parents in a recombination operator. The resulting offspring are mutated, then they are improved by using a local search operator for a fixed number of iterations L. Finally, the population is updated by applying a survival selection strategy.

The iterative process described above stops when a predefined maximum number of generations ($maxGenerations$) is reached. Algorithm 1 presents the pseudo code of IMA.

Algorithm 1. Improved Memetic Algorithm (IMA).

IMA(*A graph* $G(V, E)$)
begin
    $P \leftarrow$ initPopulation($|P|$)
    while not stopCondition() **do**
        for $i \leftarrow 1$ **to** *offspring* **do**
            // select two parents $a, b \in P$
            $(a, b) \leftarrow$ selectParents(P)
            $c \leftarrow$ recombineIndividuals(a, b)
            $c' \leftarrow$ mutation(c)
            $c'' \leftarrow$ localSearch(c', L)
            insertIndividual(c'', P)
        end
        $P \leftarrow$ updatePopulation(P)
    end
    return *The best solution found*
end

3.3   Initializing the Population

After comparing different heuristics for constructing labelings for the antibandwidth problem we have decided to use a variant of the heuristic LAH reported in [13]. Our labeling heuristic constructs a level structure of a graph using a breadth first search procedure exactly like LAH does. Then, the vertices are labeled one at a time following the order of this level structure and starting randomly either by the even or the odd levels. The main difference of our labeling heuristic with respect to LAH is that we do not select the next vertex to label in a greedy manner.

In our preliminary experiments we have found that a good balance between diversity and quality for the initial population is reached using a labeling built with our heuristic combined with $|P| - 1$ distinct randomly generated labelings.

3.4   Selection Mechanisms

In this implementation mating selection $(selectParents(P))$ prior to recombination is performed by tournament selection, while one of the following standard schemes is used for the survival selection $(updatePopulation(P))$: $(\mu + \lambda)$, (μ, λ) and (μ, λ) with elitism [14].

3.5   Recombination Operators

The recombination (crossover) operator plays a very important role in any Memetic Algorithm. Indeed, it is this operator that is responsible for creating potentially promising individuals. There are several crossover operators reported in the literature that can be applied to permutation problems [15–18].

In our computational experiments, described in Section 4.2, we compare the following three crossover operators in order to identify the most suitable one for the antibandwidth problem.

The *Cycle Crossover* (CX) operator [17], which preserves the information contained in both parents in the sense that all elements of the offspring are taken from one of the parents, i.e., CX does not perform any implicit mutation. The *Partially Matched Crossover* (PMX) operator [16], designed to preserve absolute positions from both parents. The Order Crossover (OX) operator [15], which is implemented to inherit the elements between two randomly selected crossover points, inclusive, from the first parent in the same order and position as they appeared in it. The remaining elements are inherited from the second parent in the order in which they appear in that parent, beginning with the first position following the second crossover point and skipping over all elements already present in the offspring.

3.6   Mutation Operator

In our IMA implementation the mutation operator was designed to introduce diversity into the population. It starts receiving a configuration (labeling) c produced by the recombination operator. Then, every label in c is exchanged with another randomly selected one with certain probability (mutation probability). Finally, these exchange operations allow to produce a new labeling c'.

3.7   Local Search Operator

The purpose of the local search (LS) operator $localSearch(c', L)$ is to improve a configuration c' produced by the mutation operator for a maximum of L iterations before inserting it into the population. In general, any local search method can be used. In our implementation, we have decided to use a Stochastic Hill Climbing (SHC) algorithm because it only needs as parameter the maximum number of iterations.

In our SHC-based LS operator the neighborhood $\mathcal{N}(\varphi)$ of a configuration φ is such that for each $\varphi \in \mathscr{L}$, $\varphi' \in \mathcal{N}(\varphi)$ if and only if φ' can be obtained by exchanging the labels of any pair of vertices from φ. The main advantage of this neighborhood function is that it allows an incremental fitness evaluation of the neighboring solutions.

The LS operator starts from the current solution $c' \in \mathscr{L}$ and at each iteration randomly generates a neighboring solution $c'' \in \mathcal{N}(c')$. The current solution is replaced by this neighboring solution if the fitness of c'' improves or equals that of c'. The algorithm stops when it reaches a predefined maximum number of iterations L, and returns the best labeling found.

4   Computational Experiments

In this section two main experiments accomplished to evaluate the performance of the proposed IMA algorithm and some of its components are presented. The objective of the first experiment is to determine both a component combination, and a set of parameter values which permit IMA to attain the best trade-off between solution quality and computational effort. The purpose of the second of our experiments is to carry out a performance comparison of IMA with respect to an existing state-of-the-art Memetic Algorithm called MAAMP [13].

For these experiments IMA was coded in C and compiled with *gcc* using the optimization flag -*O3*. It was run sequentially into a CPU Xeon at 2.67 GHz, 1 GB of RAM with Linux operating system. Due to the non-deterministic nature of the algorithm, 30 independent runs were executed for each of the selected benchmark instances in each experiment.

4.1   Benchmark Instances and Comparison Criteria

The test-suite that we have used in our experiments is the same proposed by Bansal and Srivastava [13]. It consists of 30 undirected planar graphs taken from the Rome set which are employed in graph drawing competitions. All of them have a number of vertices between 50 and 100. These instances are publicly available at: http://www.graphdrawing.org/data.

The criteria used for evaluating the performance of the algorithms are the same as those used in the literature: the best antibandwidth found for each instance (bigger values are better) and the expended CPU time in seconds.

4.2   Components and Parameters Tunning

Optimizing parameter settings is an important task in the context of algorithm design. Different procedures have been proposed in the literature to find the most suitable combination of parameter values [19–21]. In this paper we employ a tunning methodology, previously reported in [22], which is based on Combinatorial Interaction Testing (CIT) [7]. We have decided to use CIT, because it allows to significantly reduce the number of tests (experiments) needed to determine the best parameter settings of an algorithm. Instead of exhaustive testing all the parameter value combinations of the algorithm, it only analyzes the interactions of

Table 1. Input parameters of the IMA algorithm and their selected values

$|P|$	Cx	$ProbCx$	$ProbMuta$	$Survival$	$ProbLS$	L
40	CX	0.70	0.00	$(\mu + \lambda)$	0.05	1000
80	PMX	0.80	0.05	(μ, λ)	0.10	5000
120	OX	0.90	0.10	(μ, λ) with elitism	0.15	10000

t (or fewer) input parameters by creating interaction test-suites that include at least once all the t-way combinations between these parameters and their values.

Covering arrays (CAs) are combinatorial designs which are extensively used to represent those interaction test-suites. A covering array, $CA(N; t, k, v)$, of size N, strength t, degree k, and order v is an $N \times k$ array on v symbols such that every $N \times t$ sub-array includes, at least once, all the ordered subsets from v symbols of size t (t-tuples) [23]. The minimum N for which a $CA(N; t, k, v)$ exists is the *covering array number* and it is defined according to the following expression: $CAN(t, k, v) = min\{N : \exists CA(N; t, k, v)\}$.

CAs are used to represent an interaction test-suite as follows. In an algorithm we have k input parameters. Each of these has v values or levels. An interaction test-suite is an $N \times k$ array where each row is a test case (i.e., a covering array). Each column represents an input parameter and a value in the column is the particular configuration. This test-suite allows to cover all the t-way combinations of input parameter values at least once. Thus, the costs of tunning the algorithm can be substantially reduced by minimizing the number of test cases N in the covering array. Next, we present the details of the tunning process, based on CIT, for the particular case of our IMA algorithm.

First, we have identified $k = 7$ input parameters used for IMA: population size $|P|$, crossover operator Cx, crossover probability $ProbCx$, mutation probability $ProbMuta$, survival selection strategy $Survival$, local search probability $ProbLS$ and maximum number of local search iterations L. Based on some preliminary experiments, $v = 3$ reasonable values (shown in Table 1) were selected for each one of those input parameters.

We have constructed the smallest possible covering array $CA(40; 3, 7, 3)$, shown (transposed) in Table 2, by using the Memetic Algorithm reported in [24]. This covering array can be easily mapped into an interaction test-suite by replacing each symbol from each column to its corresponding parameter value. For instance, we can map 0 in the first column (the first line in Table 2) to $|P| = 40$, 1 to $|P| = 80$ and 2 to $|P| = 120$. The resulting interaction test-suite contains, thus, 40 test cases (parameter settings) which include at least once all the 3-way combinations between IMA's input parameters and their values[2].

Each one of those 40 test cases was used to executed 30 times the IMA algorithm over a subset of 6 representative graphs,[3] selected from the benchmark instances described in Sect. 4.1. The data generated by these 7200 executions is summarized in Fig. 1, which depicts the average antibandwidth reached by each test case over the 6 selected graphs, as well as the average CPU time expended.

---

[2] In contrast, with an exhaustive testing which contains $3^7 = 2187$ test cases.

[3] One graph for each size $50 \leq |V| \leq 100$ in the original set.

Table 2. Covering array CA$(40; 3, 7, 3)$ representing an interaction test-suite for tunning IMA (transposed)

```
0 2 2 1 0 1 1 2 2 0 0 1 2 0 0 0 1 1 1 0 2 0 2 2 2 0 2 0 0 1 1 1 2 0 1 2 1 1 2 0
1 0 2 1 2 1 1 1 1 2 0 2 0 2 0 1 0 2 0 1 1 0 0 2 1 0 2 0 1 0 1 0 1 2 2 0 2 2 2 0
1 0 0 2 0 2 0 2 0 2 1 2 2 0 1 2 1 1 0 0 1 2 1 2 0 1 2 0 2 0 1 2 1 1 0 0 2 1 1 0
0 2 0 1 2 0 0 2 1 0 2 2 0 1 0 1 0 0 2 2 0 1 2 1 1 1 0 0 2 1 2 2 1 2 1 1 0 1 2 0
0 2 0 0 1 2 1 0 1 1 0 1 1 0 1 1 0 2 0 0 2 0 1 2 2 2 1 2 2 1 1 2 0 2 2 0 0 1 0 1
0 1 0 2 2 0 1 0 0 0 1 1 1 1 1 2 2 1 0 2 2 0 0 1 0 2 2 0 1 1 2 2 1 0 2 2 1 0 2 2
2 0 0 2 1 0 1 1 0 2 1 0 2 2 0 1 0 2 2 0 1 0 1 1 2 2 0 1 2 2 2 1 0 0 0 1 1 1 2 2
```

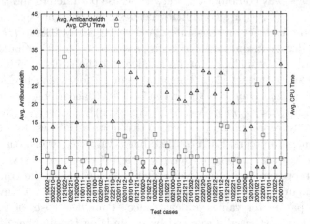

Fig. 1. Average results obtained in the tuning experiments using 40 parameter value combinations over a subset of 6 representative graphs

From this graphic we have selected the 5 test cases which yield the best results. Their average antibandwidth and the average CPU time in seconds are presented in Table 3. This table allowed us to observe that the parameter setting giving the best trade-off between solution quality and computational effort corresponds to the test case number 40 (shown in bold). The best average antibandwidth with an acceptable speed is reached with the following input parameter values: population size $|P| = 40$, Cycle Crossover (CX) operator, crossover probability $ProbCx = 0.90$, mutation probability $ProbMuta = 0.00$, (μ, λ) survival selection strategy, local search probability $ProbLS = 0.15$ and maximum number of local search iterations $L = 10000$. These values are thus used in the experimentation reported in the next section.

Table 3. Results from the 5 best parameter test cases in the tuning experiments

Num.	Test case	Avg. antibandwidth	Avg. CPU time
13	2020112	31.527	11.802
40	**0000122**	**31.011**	**5.209**
10	0220102	30.616	2.112
7	1100111	30.511	4.611
27	2220120	29.183	2.223

4.3   Comparison between IMA and MAAMP

In this experiment a performance comparison of the best bounds achieved by IMA with respect to those produced by the MAAMP algorithm [13] was carried out over the test-suite described in Sect. 4.1.

Table 4 displays the detailed computational results produced by this experiment. The first three columns in the table indicate the name of the graph as well as its number of vertices and edges. The theoretical upper bound (C^*) reported in [13] for those graphs is presented in Column 4. The best (C) and average $(Avg.)$ antibandwidth attained by MAAMP in 10 executions and its average CPU time in seconds are listed in columns 5 to 7. These results were taken directly from [13], where a Pentium 4 at 3.2 GHz and 1 GB of RAM system was used to execute the algorithm. Next four columns provide the best (C), average $(Avg.)$ and standard deviation $(Dev.)$ of the antibandwidth found by IMA over 30 independent executions and the average CPU time (T) in seconds expended. The running times from MAAMP and IMA cannot be directly compared because they were executed on different computational platforms. Nevertheless, we have scaled, by a factor of 2.71, our execution times according to the Standard Performance Evaluation Corporation[4] in order to present them in a normalized form in Column 12 $(\overline{T})$. Finally, the difference (Δ_C) between the best result produced by our IMA algorithm and that achieved by MAAMP is depicted in the last column.

Analyzing the data presented in Table 4 lead us to the following main observations. First, the solution quality attained by the proposed IMA algorithm is very competitive with respect to that produced by the existing Memetic Algorithm called MAAMP [13], since IMA provides solutions whose costs (antibandwidth) are closer to the theoretical upper bounds (compare Columns 4, 5 and 8). Indeed, IMA consistently improves the best antibandwidth found by MAAMP, obtaining an average amelioration of $\Delta_C = -4.93$.

Second, one observes that for the selected instances the antibandwidth found by IMA presents a relatively small standard deviation (see Column $Dev.$). It is an indicator of the algorithm's precision and robustness since it shows that in average the performance of IMA does not present important fluctuations.

Third, we can notice that MAAMP is the most time-consuming algorithm. It uses an average of 257.79 seconds for solving the 30 selected instances, while IMA employs only 25.82 seconds (see column $\overline{T}$).

The outstanding results achieved by IMA are better illustrated in Fig. 2. The plot represents the studied instances (ordinate) against the best solution (antibandwidth) attained by the compared algorithms (abscissa). The theoretical upper bounds for these graphs are shown with squares, the previous best-known solutions provided by MAAMP [13] are depicted as circles, while the bounds computed with our IMA algorithm are shown as triangles. From this figure it can be seen that IMA consistently outperforms MAAMP, achieving for certain instances, like the graph $ug5$-100, an important increase in solution cost (Δ_C up to -11).

Thus, as this experiment confirms, our IMA algorithm is more effective than the existing Memetic Algorithm, called MAAMP.

---

[4] http://www.spec.org

Table 4. Performance comparison between MAAMP and IMA over 30 undirected planar graphs from the Rome set

Graph	$|V|$	$|E|$	C^*	MAAMP			IMA					Δ_C
				C	Avg.	T	C	Avg.	Dev.	T	$\overline{T}$	
ug1-50		64		18	17.40	65.28	21	20.97	0.03	8.51	23.07	-3
ug2-50		66		18	16.80	33.51	21	20.00	0.21	2.38	6.46	-3
ug3-50	50	63	24	17	16.80	52.04	21	21.00	0.00	3.61	9.77	-4
ug4-50		70		19	18.40	59.55	21	20.37	0.52	2.74	7.42	-2
ug5-50		62		20	19.80	83.44	22	21.40	0.25	12.33	33.41	-2
ug1-60		79		21	20.20	155.98	25	24.13	0.12	3.59	9.72	-4
ug2-60		81		23	21.80	21.80	24	23.53	0.26	2.41	6.54	-1
ug3-60	60	84	29	22	20.20	116.12	24	23.33	0.30	7.27	19.71	-2
ug4-60		79		22	21.20	235.20	25	24.27	0.20	6.32	17.12	-3
ug5-60		80		21	20.80	70.20	25	23.87	0.19	3.26	8.84	-4
ug1-70		98		26	23.80	354.52	28	26.83	0.35	6.13	16.60	-2
ug2-70		82		28	15.80	54.20	32	31.30	0.22	9.68	26.22	-4
ug3-70	70	82	34	27	26.20	116.31	31	30.20	0.17	5.72	15.49	-4
ug4-70		97		23	22.60	74.50	29	28.07	0.06	15.42	41.77	-6
ug5-70		88		26	25.50	323.45	29	28.47	0.26	13.03	35.31	-3
ug1-80		92		34	33.60	61.07	37	36.13	0.26	9.31	25.24	-3
ug2-80		93		31	30.20	226.90	35	34.07	0.34	14.26	38.64	-4
ug3-80	80	95	39	28	27.20	138.68	35	34.23	0.46	9.72	26.35	-7
ug4-80		101		27	25.20	582.50	34	32.80	0.23	11.05	29.95	-7
ug5-80		94		27	25.80	190.52	34	33.10	0.51	10.74	29.11	-7
ug1-90		102		32	29.80	150.40	39	38.27	0.41	8.78	23.78	-7
ug2-90		114		35	34.60	469.62	38	36.73	0.34	7.90	21.41	-3
ug3-90	90	108	44	31	29.80	316.29	40	38.83	0.49	16.84	45.64	-9
ug4-90		99		34	33.30	381.20	40	39.23	0.25	11.77	31.90	-6
ug5-90		104		34	31.80	815.47	39	38.50	0.47	10.50	28.46	-5
ug1-100		114		34	32.20	499.20	44	43.07	0.34	17.61	47.71	-10
ug2-100		114		40	38.40	715.90	44	42.80	0.23	11.22	30.40	-4
ug3-100	100	116	49	33	31.60	256.40	43	42.33	0.57	15.10	40.92	-10
ug4-100		122		35	33.80	419.80	43	42.13	0.33	15.23	41.27	-8
ug5-100		125		31	30.60	693.65	42	40.60	1.14	13.41	36.33	-11
Avg.				27.23		257.79	32.17		0.32	9.53	25.82	-4.93

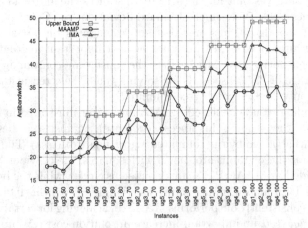

Fig. 2. Performance comparison between MAAMP and IMA with respect to the theoretical upper bounds

5 Conclusions and Further Work

In this paper, an Improved Memetic Algorithm (IMA) designed to compute near-optimal solutions for the antibandwidth problem was presented. IMA's components and parameter values were carefully determined, through the use of a tunning methodology based on Combinatorial Interaction Testing [7], to yield the best solution quality in a reasonable computational time.

The practical usefulness of this fine-tunned IMA algorithm was assessed with respect to an existing state-of-the-art Memetic Algorithm, called MAAMP [13] over a set of 30 well-known benchmark graphs taken from the literature. The results show that our IMA algorithm was able to consistently produce labelings with higher antibandwidth values than those furnished by MAAMP. Furthermore, IMA achieves those results by employing only a small fraction (10.01%) of the total time used by MAAMP.

This work opens up a range of possibilities for future research. Currently we are interested in solving the antibandwidth problem for bigger graphs with different topologies efficiently. To accomplish it we would like to improve the Memetic Algorithm presented here by implementing some of the algorithmic extensions exposed in [25, 26], like self-adaptation and parametric adaptation.

References

1. Leung, J., Vornberger, O., Witthoff, J.: On some variants of the bandwidth minimization problem. SIAM Journal on Computing 13(3), 650–667 (1984)
2. Yixun, L., Jinjiang, Y.: The dual bandwidth problem for graphs. Journal of Zhengzhou University 35(1), 1–5 (2003)
3. Hale, W.K.: Frequency assignment: Theory and applications. Proceedings of the IEEE 68(12), 1497–1514 (1980)
4. Roberts, F.S.: New directions in graph theory. Annals of Discrete Mathematics 55, 13–44 (1993)
5. Cappanera, P.: A survey on obnoxious facility location problems. Technical report, Uni. di Pisa (1999)
6. Burkard, R.E., Donnani, H., Lin, Y., Rote, G.: The obnoxious center problem on a tree. SIAM Journal on Computing 14(4), 498–509 (2001)
7. Cohen, D.M., Dalal, S.R., Parelius, J., Patton, G.C.: The combinatorial design approach to automatic test generation. IEEE Software 13(5), 83–88 (1996)
8. Miller, Z., Pritikin, D.: On the separation number of a graph. Networks 19, 651–666 (1989)
9. Yao, W., Ju, Z., Xiaoxu, L.: Dual bandwidth of some special trees. Journal of Zhengzhou University Natural Science Edition 35, 16–19 (2003)
10. Calamoneri, T., Missini, A., Török, L., Vrt'o, I.: Antibandwidth of complete k-ary trees. Electronic Notes in Discrete Mathematics 24, 259–266 (2006)
11. Török, L.: Antibandwidth of three-dimensional meshes. Electronic Notes in Discrete Mathematics 28, 161–167 (2007)
12. Raspaud, A., Schröder, H., Sykora, O., Török, L., Vrt'o, I.: Antibandwidth and cyclic antibandwidth of meshes and hypercubes. Discrete Mathematics 309(11), 3541–3552 (2009)

13. Bansal, R., Srivastava, K.: Memetic algorithm for the antibandwidth maximization problem. Journal of Heuristics 17(1), 39–60 (2011)
14. Eiben, A.E., Smith, J.E.: Introduction to Evolutionary Computing, 1st edn. Springer (2007)
15. Davis, L.: Applying adaptive algorithms to epistatic domains. In: Proceedings of the International Joint Conference on Artificial Intelligence, pp. 162–164. Morgan Kaufmann (1985)
16. Goldberg, D.E., Lingle, R.: Alleles, loci, and the travelling salesman problem. In: Proceedings of the 1st International Conference on Genetic Algorithms and their Applications, pp. 154–159. Lawrence Erlbaum Associates (1985)
17. Oliver, I.M., Smith, D.J., Holland, J.R.C.: A study of permutation crossover operators on the travelling salesman problem. In: Proceedings of the 2nd International Conference on Genetic Algorithms and their Applications, pp. 224–230. Lawrence Erlbaum Associates (1987)
18. Freisleben, B., Merz, P.: A genetic local search algorithm for solving symmetric and asymmetric traveling salesman problems. In: Proceedings of IEEE International Conference on Evolutionary Computation, pp. 616–621. IEEE Press (1996)
19. Adenso-Diaz, B., Laguna, M.: Fine-tuning of algorithms using fractional experimental design and local search. Operations Research 54(1), 99–114 (2006)
20. de Landgraaf, W.A., Eiben, A.E., Nannen, V.: Parameter calibration using meta-algorithms. In: Proceedings of the IEEE Congress on Evolutionary Computation, pp. 71–78. IEEE Press (2007)
21. Gunawan, A., Lau, H.C., Lindawati: Fine-Tuning Algorithm Parameters Using the Design of Experiments Approach. In: Coello, C.A.C. (ed.) LION 2011. LNCS, vol. 6683, pp. 278–292. Springer, Heidelberg (2011)
22. Gonzalez-Hernandez, L., Torres-Jimenez, J.: MiTS: A New Approach of Tabu Search for Constructing Mixed Covering Arrays. In: Sidorov, G., Hernández Aguirre, A., Reyes García, C.A. (eds.) MICAI 2010, Part II. LNCS (LNAI), vol. 6438, pp. 382–393. Springer, Heidelberg (2010)
23. Colbourn, C.J.: Combinatorial aspects of covering arrays. Le Matematiche 58, 121–167 (2004)
24. Rodriguez-Tello, E., Torres-Jimenez, J.: Memetic Algorithms for Constructing Binary Covering Arrays of Strength Three. In: Collet, P., Monmarché, N., Legrand, P., Schoenauer, M., Lutton, E. (eds.) EA 2009. LNCS, vol. 5975, pp. 86–97. Springer, Heidelberg (2010)
25. Moscato, P., Berretta, R., Cotta, C.: Memetic algorithms. In: Wiley Encyclopedia of Operations Research and Management Science. John Wiley & Sons, Inc. (2011)
26. Neri, F., Cotta, C., Moscato, P. (eds.): Handbook of Memetic Algorithms. SCI, vol. 379. Springer, Heidelberg (2012)

Adaptive Play in a Pollution Bargaining Game

Vincent van der Goes

Vrije Universiteit Amsterdam, The Netherlands
vgoes@feweb.vu.nl

Abstract. We apply adaptive play to a simplified pollution game with
two players. We find that agents with longer memory paradoxically perform worse in the long run. We interpret this result as an indication that
adaptive play may be too restrictive as a model of agent behaviour in
this context, although it can serve as a starting point for further research
on bounded rationality in pollution games.

Keywords: evolutionary economics, game theory, adaptive play.

1 Introduction

Global environmental problems, such as pollution, provide real-world examples
of a *social dilemma*. Countries involved in an environmental issue could often
benefit from mutual cooperation. However, if they choose to do so, any country
in the cooperation might benefit from *free riding*. That is, each country could
possibly improve net welfare by defecting and profiting from the efforts at reducing pollution, without reducing its own pollution.

So it is in the mutual interest of countries involved in such an environmental
problem to find a way to facilitate cooperation. To this end, countries have signed
treaties known as *International Environmental Agreements* (IEA's).

In particular, the case of greenhouse gas emissions is receiving a lot of attention today. Greenhouse gas emission is a form of *transboundary pollution*, i.e. the
impacts of emissions are global and do not depend on the location where they
originate. As reduction of greenhouse gas emissions is very costly, it provides a
good example of a social dilemma. Agreeing on an appropriate IEA has proven
difficult in practice.

In the literature this has sparked a discussion about the question why negotiations on reduction of emissions are so difficult and what kind of policy
instruments could help on reaching an agreement. For a recent overview of the
literature in this area we refer the reader to [2] and [3].

A common way of modelling emission reductions is as a repeated game. The
dynamics of a realistic model are highly complex, as there are several complicat-
ing factors at play. The game changes over time, as greenhouse gas concentrations
build up and new technologies become available.

Also, there is a high degree of uncertainty, especially considering the long time
scale at which the climate responds [6]. First of all *systematic uncertainty*: damage as a function of emissions is hard to predict. Secondly *strategic uncertainty*:
agents can not predict the behaviour of other agents very well.

J.-K. Hao et al. (Eds.): EA 2011, LNCS 7401, pp. 133–144, 2012.
© Springer-Verlag Berlin Heidelberg 2012

In this paper we explore the possibility of modelling bargaining for reduction of emissions of a transboundary pollutant, while assuming only bounded rationality and imperfect information. We focus on the question whether agents will be able to coordinate on an efficient outcome under these conditions. In particular, we look at the effect of asymmetry in information.

The agents in our model will bargain on pollution abatement levels, using the *adaptive play* mechanism [9] to optimize their results. In adaptive play, agents are *myopic*: payoff in future rounds is not taken into account. Instead, the agents remember the earlier choices of other agents, assume they will make similar choices at present and search for a best response. The agents will also make occasional errors. This is to account for the fact that agents in the real world will occasionally act irrational, or there might be an unexpected external factor playing a role.

The bargaining game in our model is defined in two steps. The first step in the definition of our model is a toy model of transboundary pollution, the *pollution game*. The model is attractive for study and has been widely used for analysis with perfect information and rational agents, e.g. [1] [8]. We make a further simplification to the model, as we limit it to two agents.

Within the context of this pollution game, we assume that agents can make binding agreements on abatement for one round. At the beginning of each round, the agents will play a bargaining game over their abatement levels during that round. In the bargaining game both agents simultaneously make a proposal for the abatement levels. If the two proposals are mutually compatible, an agreement is made for that round. If the proposals are not compatible with each other, the agents have failed to make an agreement in that round and they will both choose their abatement levels according to the Nash equilibrium of the pollution game, instead.

We investigate the behavior of this model by running simulations. We find that agents are able to reach Pareto efficient outcomes, even though this is not true for all parameter settings. However, when one agent has a longer memory of past actions than the other, our behaviour model predicts that he will be worse off in the long run. This is a rather paradoxal finding. There is very little difference between our model and the bargaining model in [10], except that the underlying pollution game has a different payoff structure. Yet, in that study agents with longer memory were consistently better off.

This leads to two conclusions. First, it shows that the result in [10] does not generalize very well to other classes of bargaining games. Secondly, it is an indication that adaptive play may compromise the rationality of agents too strongly to be suitable as a model for this particular bargaining game.

The remainder of this paper is organised as follows. Section 2 describes the bargaining game. The pollution game is defined and analyzed in section 2.1 and in section 2.2 it is extended with a bargaining mechanism. Section 2.3 defines the behaviour of the agents. Section 3 contains the experiments. The simulations are motivated and detailed in section 3.1. The experimental results are given in section 3.2 and section 4 concludes and suggests paths for further research.

2 Model

2.1 The Pollution Game

We model negotiations between two agents on emissions of a pollutant. The pollutant is transboundary and damage functions depend only on total emissions. Agents have the options to *abate*, lowering their emission for one period at a cost.

The emphasis of the study is on learning dynamics. For this purpose we employ a strongly simplified pollution model, known as the standard model [1] [8]. We assume that costs of abatement are independent between agents and do not change over time. Further, it is assumed that the pollutant does not accumulate in the environment, even though greenhouse gases are in reality a stock pollutant. Also, all agents have the same damage function.

Agents are labelled $i = 1, 2$. They have an fixed unabated emission level $a/2$, where a is the aggragate emission level. They can spend on abatement, resulting in benefits for every agent. Hence, the benefits to agent i depend on the total abatement level Q. We assume linear marginal abatement benefits $B_i(Q)$. This leads to a quadratical benefit function, as specified in equation 1.

$$B_i(Q) = b(aQ - Q^2/2)/2 \tag{1}$$

The marginal benefit of the first unit of abatement is $ba/2$, with b a positive parameter. When total abatement equals total emission, the marginal benefit of further abatement is zero.

The costs of abatement $C_i(q_i)$ for agent i depends only on the abatement q_i of that agent, where $Q = q_1 + q_2$. The marginal costs of abatement are assumed to rise linearly, leading to the quadratic cost function in equation 2.

$$C_i(q_i) = cq_i^2/2 \tag{2}$$

where c is a positive parameter. The resulting net payoff vector or *utility* vector is shown in equation 3.

$$u_i(q_1, q_2) = B_i(q_1 + q_2) - C_i(q_i) \tag{3}$$

The Nash Equilibrium. In the *Nash equilibrium*, both agents optimize their own payoff, without considering the possibility of mutual benefits from cooperation. This is known as *non-cooperative* behaviour.

In the pollution game of section 2.1, the Nash equilibrium can be calculated by taking the partial derivatives of the utility functions of both agents with respect to their own abatement level. Solving for the first order conditions yields a single symmetric Nash equilibrium, given in equation 4.

$$q_i = q_D = \frac{a}{2(1 + c/b)} \tag{4}$$

Here the Nash equilibrium is denoted as q_D, because it will serve as the *disagreement point* in the bargaining game in section 2.2. The Nash equilibrium yields equal utility $u_D = u_1(q_D, q_D)$ to both players.

The Pareto Frontier. The Nash equilibrium of this game is not Pareto optimal. Both agents could increase their payoff by cooperation. Therefore, this game is an example of a *social dilemma*, similar to the *prisoner's dilemma*.

If the agents cooperate, they will ideally choose Pareto optimal abatement levels. Additionally, both agents should receive greater payoff than in the Nash equilibrium. The symmetric solution on the Pareto frontier, the *Nash bargaining solution* (q^*, q^*) (see equation 5), satisfies both conditions. In [7], it is shown that under certain assumptions, the Nash bargaining solution will be the result of negotiations between two agents equal bargaining power.

$$q^* = \frac{a}{2 + c/b} \tag{5}$$

However, it should be noted that in our simulations the agents will not have perfect information, nor will they necessarily have identical bargaining power. The complete Pareto frontier can be characterized by maximizing the weighted average of u_1 and u_2 in equation 6, with weight α.

$$\alpha u_1(q_1, q_2) + (1 - \alpha)u_2(q_1, q_2) \tag{6}$$

The function in equation 6 has a single maximum at $(q_1^*(\alpha), q_2^*(\alpha))$ in 7.

$$q_1^*(\alpha) = \frac{(1 - \alpha)a}{1 + 2\alpha(1 - \alpha)c/b}$$
$$q_2^*(\alpha) = \frac{\alpha a}{1 + 2\alpha(1 - \alpha)c/b}$$

$$\tag{7}$$

In the special case that $\alpha = 1/2$, the solution in equation 7 reduces to the Nash bargaining solution (q^*, q^*). The extreme points of the Pareto front are the cases $\alpha = 0$ and $\alpha = 1$. In these cases, one agent does not abate and the other agent chooses an abatement level of a.

2.2 The Bargaining Game

The pollution game described in section 2.1 is assumed to be repeated for an unlimited number of rounds. Under such conditions, cooperation is possible, according to the folk theorem [5]. However, the incentive for agents to cooperate, rather than to play the Nash equilibrium, is that cooperation in this round may be rewarded by the other agent in later rounds.

In this work we will look at agents with no forward looking abilities. With such limited rationality, agents will not be able to cooperate unless some bargaining mechanism is added to the game.

Therefore, we add a simple bargaining system. Let agent $-i$ denote the opposite agent from agent i. Instead of choosing abatement levels directly, both agents simultaneously make a proposal $p_i^t = (\hat{q}_{i,1}^t, \hat{q}_{i,2}^t)$ for the abatement levels in round t. Here $\hat{q}_{i,i}^t$ is the level of abatement that agent i is offering himself. However, in exchange, agent i demands an abatement level of at least $\hat{q}_{i,-i}$ from agent $-i$. Only if the offers and demands of both agents are compatible, the negotiations result in cooperation for round t. Otherwise, the agents fail to reach an agreement in round t and they will play the Nash equilibrium of the pollution game instead.

Denote the boolean value of this requirement as $\Delta \in [\textbf{true}, \textbf{false}]$:

$$\Delta(p_i^t, p_{-i}^t) = \forall i : \hat{q}_{i,i}^t \geq \hat{q}_{-i,i}^t \tag{8}$$

If condition 8 is met, both agents are willing to abate at least as much as the other agent demands. But it is not yet clear what the final agreement will be. Will q_i^t be set to the level $\hat{q}_{i,i}^t$ that i offered himself, or to the level $\hat{q}_{-i,i}^t$ that the other agent demanded? In general, we could imagine the final result of negotiations to be any value in between the two. We assume that the agents will share the difference in a manner agreed upon prior to negotiations. The final abatement level of agent i will be a weighted average of the abatement level that agent i offered and the abatement level that was required, as in equation 9. The effect of varying λ will be analysed in sections 3 and 4.

$$q_i^t = q_i(p_i^t, p_{-i}^t) = \begin{cases} \lambda \hat{q}_{i,i}^t + (1 - \lambda)\hat{q}_{-i,i}^t & \text{if } \Delta(p_i^t, p_{-i}^t) \\ q_D & \text{otherwise} \end{cases} \tag{9}$$

Note that if requirement 8 fails, the agents abort negotiations and instead revert to the disagreement point, which is the Nash equilibrium of the pollution game.

The final utility level of agent i in round t, $u_i^t(q_i^t, q_{-i}^t)$, then follows by inserting these values into equation 3:

$$u_i^t = u_i(q_i^t, q_{-i}^t) = B_i(q_i^t, q_{-i}^t) - C_i(q_i^t) \tag{10}$$

Nash Equilibria of the Bargaining Game. The negotiation process thus defined constitutes a new game, where the actions agents have to choose from are the proposals they make. In order to distinguish this game from the pollution game itself, we will refer to it as the *bargaining game*. Where the pollution game has only a single Nash equilibrium, the bargaining game has many.

Consider the solution where, for $i = 1, 2$, $\hat{q}_{i,i}^t = 0$ and $\hat{q}_{i,-i}^t = a$. This is a degenerate Nash equilibrium. Since requirement 8 is not met, the result is that both players play the Nash equilibrium $q_i^t = q_D$ in the pollution game. Neither agent can improve his own utility by deviating unilaterally.

The game also has many non-degenerate equilibria, where both players end up with higher utility than in the Nash equilibrium of the pollution game. For example, consider $\hat{q}^t_{i,i} = \hat{q}^t_{i,-i} = q^*$. This leads to the Nash Bargaining Solution of the pollution game. Both agents have improved their utility over u_D and neither agent could improve his utility further by a unilateral deviation from his negotiation strategy.

Any non-degenerate equilibrium of the bargaining game must be Pareto dominant over the Nash equilibrium of the pollution game. If it does not, then at least one agent is worse off than u_D and therefore has an incentive to let the negotiations fail instead, by lowering his abatement offer.

If $\lambda < 1$, any non-degenerate Nash equilibrium, with at least one player ending up with a higher utility than u_D, must have $\hat{q}^t_{i,i} = \hat{q}^t_{-i,i}$. If $\hat{q}^t_{i,i} < \hat{q}^t_{-i,i}$ then the negotiations fail. But if $\hat{q}^t_{i,i} > \hat{q}^t_{-i,i}$, then agent $-i$ has an incentive to increase $\hat{q}^t_{-i,i}$, since by equation 9 this would increase the abatement of i, but leave his own abatement unchanged. As an increase in the abatement of i increases the benefits for $-i$ but not his costs, the utility of agent $-i$ would increase.

Provided that $0 < \lambda < 1$, any solution of the bargaining game that satisfies both of these conditions is a Nash equilibrium of the bargaining game. Neither player has an incentive to change his own proposed abatement $\hat{q}^t_{i,i}$, because it would either let the negotiations fail or it would cause him to increase his abatement level further above q_D, lowering his utility. Similarly, neither player has an incentive to change the abatement level he demands from the opponent, $\hat{q}^t_{i,-i}$.

Now that we identified the non-degenerate Nash equilibria (at least for $0 < \lambda < 1$), we can calculate the corresponding range of values for q^t_i.

The extreme points of the region that Pareto dominates q_D are on the Pareto frontier of the pollution game, as defined by $q^*_i(\alpha)$ in equation 7. We have the constraint that $q^t_i \leq q_D$. Solving for α in equation 7, we find that the region of non-degenerate Nash equilibria is bounded by the values q_{min} and q_{max} in equation 11.

We will assume that agents only consider values for their proposals in the range $[q_{min}, q_{max}]$. It could be argued that the agents have insufficient information to know q_{min} and q_{max} ahead of time. However, if our agent model is applied to the Pollution Game itself, rather than to the Bargaining Game, it will converge to the unique Nash equilibrium (q_D, q_D), so it is not altogether unreasonable to postulate that agents are aware of $q_{min} = q_D$, at least.

$$\alpha_{min} = \frac{\sqrt{b^2 + 2bc} - b}{2c}$$
$$\alpha_{max} = 1 - \alpha_{min}$$
$$q_{min} = q_D$$
$$q_{max} = q_D * \frac{\alpha_{max}}{\alpha_{min}} \qquad (11)$$

2.3 Adaptive Play

In this section we describe how agents decide what proposals to make in the negotiations of section 2.2. Agents are assumed to have bounded rationality. They do not have any forward looking abilities and do not know each others payoff functions. Instead, they form a model of the behaviour of the other agents, in order to find the best response to their actions.

We apply the *adaptive play* mechanism [9]. In adaptive play, agents assume that the other agents draw their proposals from a fixed probability distribution. In other words, they assume that the other agent plays a mixed strategy that doesn't change over time.

Agents have a limited memory m_i. Only actions chosen by the other agent in the latest m_i rounds are remembered. The agents then assume that the other agent will play any of the actions that he played during the last m_i rounds, each with equal probability.

Based on this internal model of the opponent, the agents search for an optimal response. They will maximize the expected final utility of their proposal p_i^t. Let (x, y) be a candidate proposal for p_i^t. It is evaluated by the expected value of the utility that it could yield:

$$f_i^t(x,y) = \frac{1}{m_i} \sum_{t'=t-1}^{t-m_i} u_i(q_i((x,y),p_{-i}^{t'}), q_{-i}((x,y),p_{-i}^{t'})) \qquad (12)$$

Where q_i is calculated as in equation 9 and u_i as in equation 3. The function $f_i^t(x,y)$ serves as a fitness function.

The agent will then make the proposal that maximizes this fitness function:

$$p_i^t = \operatorname*{argmax}_{(x,y)\in[q_{min},q_{max}]^2} f_i^t(x,y) \qquad (13)$$

In case there are multiple values of (x, y) that maximize $f(x, y)$, one value is drawn from the set with a uniform random distribution.

However, in adaptive play, agents are allowed to make occasional errors. These are introduced to take into account that agents in the real world occasionally behave unpredictably, because of factors external to the game. With a small probability ε, equation 13 is ignored and p_i^t is instead drawn at random from $[q_{min}, q_{max}]^2$ with uniform distribution. This is known as a mutation.

Of course, equation 12 is undefined during the first few rounds of play. For $t = 1, \ldots, \max(m_1, m_2)$, p_i^t is again drawn at random from $[q_{min}, q_{max}]^2$ with uniform distribution.

2.4 Stochastic Stability

As shown in section 2.2, the bargaining game has many Nash equilibria. It is easily verified that all of the non-degenerate Nash equilibria of the game are also evolutionary stable strategies. Hence, the question is how to define or identify the most likely outcome of the game.

We are interested in finding the *stochastically stable strategies* of the game. Stochastically stable strategies are a refinement of the concept of evolutionary stable strategies. Any stochastically stable strategy is also an evolutionary stable strategy, but the reverse is not true.

In an evolutionary system such as our multi-agent model, an evolutionary stable strategy is stable in the short run, as neither agent has an incentive to deviate unilaterally. However, over increasingly long time scales there is an increasing probability that mutations disturb the equilibrium and cause the system to enter a different equilibrium. On sufficiently long timescales, the system behaves ergodic, switching back and forth between different evolutionary stable strategies. The timescale at which this kind of behaviour is typical is called the *long run*, or sometimes the *ultra-long run* [10].

Stochastically stable strategies are the evolutionary stable strategies that occur the most frequently in the long run, when the mutation rate ε approaches zero. For a full definition we refer the reader to [4].

3   Simulations

We will use simulations to gain insight into the stochastically stable strategies under different parameter settings of the model. The model will be run for a large but finite number of rounds T_{MAX}, under different parameter settings. We will then analyze the results, using autocorrelation in the series under different time lags to find empirically what timescale constitutes the long run, i.e., at what timescale the system behaves ergodic.

3.1   Experimental Setup

The primary question that we wish to address is the role of information: what happens when one agent has an information advantage over the other? What if one agent has a longer memory? We have set up a series of five experiments (E_3 and $E_6 \ldots E_9$) to compare the behaviour of the system under varying differences in memory.

The second question we wish to address is: how sensitive are the outcomes to a change in λ? For this purpose, we have set up a second series of experiments, $E_1 \ldots E_5$, where λ was varied. Experiment E_3 serves as the benchmark, as it has equal memory for both players and a value of λ of 0.5, splitting the remainder in equation 9 equally.

Table 1 provides an overview of all the parameter settings that have been used in the experiments. The ratio of b and c has been chosen such that it maximizes the potential gain from cooperation [1]. The mutation rate ε has been chosen as small as feasible to approximate the stochastically stable equilibria.

Table 1. Overview of Parameter Values

parameter	experiment								
	E_1	E_2	E_3	E_4	E_5	E_6	E_7	E_8	E_9
a	1								
b	1.64								
c	1								
ε	0.01								
T_{MAX}	10^7								
λ	0	0.25	0.5	0.75	1	0.5	0.5	0.5	0.5
m_1	8								
m_2	8	8	8	8	8	6	4	2	1

3.2   Results

In this section we will analyze the output, series of resulting utility values u_i^t. They are not time independent, but strongly autocorrelated. That is also to be expected, since a stochastic equilibrium is defined as the equilibrium that the is dominant in the long run. In order to find out what the long run is in this game, we need to find out at what time lag the autocorrelation between subsequent proposals approaches zero.

Autocorrelation can be expressed as a function of timelag, using formula 14:

$$R_i(\tau) = \frac{E[(u_i^t - \bar{u}_i)(u_i^{t+\tau} - \bar{u}_i)]}{\sigma^2} \qquad (14)$$

In figure 1 all the autocorrelation plots of the experiments are shown. As expected, autocorrelation vanishes at sufficiently long timescale. From the figure we estimate a time lag of $\tau = 50.000$ as the typical long run.

Now we can sample measurements from each of the series at intervals of 50.000 rounds and treat them as approximately independent. Figure 2 shows a histogram of the resulting sample for benchmark experiment E_3.

As can be seen from figure 2, the measurements follow a roughly bell shaped curve, with a group of outliers at the disagreement utility U_D. The outliers result from rounds where the negotiations failed. Since the interval $[q_{min}, q_{max}]$ does not contain degenerate equilibria, failed negotiations are not part of a stochastically stable equilibrium. Therefore, we will ignore those measurements in the remainder of the analysis.

We assume by the bell shaped form of the rest of the distribution that there is a single stochastically stable equilibrium and estimate it by taking the mean value. The mean values for each experiment, after dismissing outliers, can be found in table 2.

In the benchmark experiment E_3, with equal memory and $\lambda = 0.5$, agents have on average been following the Nash Bargaining Solution ($u^* = 0.3141$) nearly exactly. Indeed, a t-test shows no statistically significant difference from a normal distribution with mean u^*.

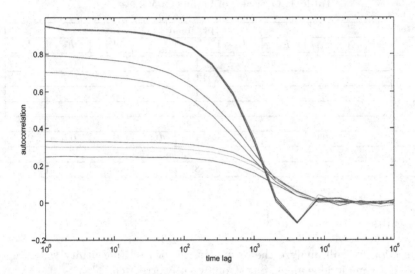

Fig. 1. Autocorrelation as a function of timelag, for both agents in each of the experiments

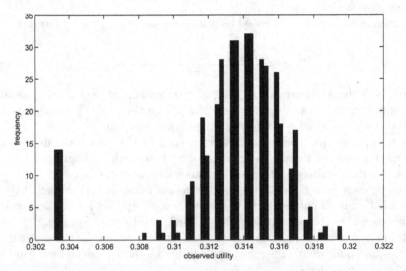

Fig. 2. Histogram of utility values sampled from E_3

When λ differs from 0.5, as in expirements E_1, E_2, E_4 and E_5, it clearly negatively affects the results of negotiations. For each of those experiments, average utility remained significantly below u^*, all at p-values below 10^{-5}.

Agents with shorter memory length have a clear advantage in the long run. Experiments E_6 to E_9 all seem to favour the less informed agent 2. However, the results from those experiments show a wide spread, with long tails. This

Table 2. Average observed utility values (outliers removed)

	E_1	E_2	E_3	E_4	E_5	E_6	E_7	E_8	E_9
λ	0	0.25	0.5	0.75	1	0.5	0.5	0.5	0.5
m_2	8	8	8	8	8	6	4	2	1
u_1	0.3098	0.3129	0.3140	0.3131	0.3101	0.3125	0.3125	0.3131	0.3129
u_2	0.3099	0.3128	0.3142	0.3133	0.3102	0.3149	0.3148	0.3141	0.3141

might explain why a t-test shows no statistical significant difference between both agents in E_8 and E_9. In E_6 and E_7, however, the difference is significant at the 1% confidence level.

The average utilities observed in the experiments with asymmetric information are all below the Pareto frontier. For example, in E_6 agent 1 had an average utility of 0.3125. It follows from 7 that a pareto optimal solution with this u_1 should have $u_2 = 0.3157$. The actual average u_2 was lower, even though a t-test shows no statistal significance. The same holds for E_7 to E_9.

4 Conclusions and Outlook

Even though our model is strongly simplified, especially as it is a two agent model, it nonetheless offers important insights into myopic bargaining on pollution games.

As often in adaptive play, agents are able to coordinate on a Pareto efficient outcome, without knowing each others payoff functions. However, this is not guaranteed and depends on parameter values of the model, moreso than in many other applications of adaptive play.

Surprisingly, it turns out that having a shorter memory is an advantage in the long run. This must be due to the structure of the pollution game, as adaptive play has been applied in this fashion to a similar bargaining game [10], where adaptive play favours the agents remembering more of the past under any parameter settings.

This result indicates that adaptive play may be too limiting for studying pollution games. While we believe that evolutionary economics is generally a promising venue for studying bounded rationality, the challenge for future research is to understand the behaviour of agents that are imperfect, yet less short sighted than in adaptive play.

Acknowledgements. The author would like to thank Cees Withagen for reviewing the manuscript. Also, thanks to Guszti Eiben and Daan van Soest for their helpful suggestions.

References

1. Barrett, S.: Self-Enforcing International Environmental Agreements. Oxford Economic Papers 46, 804–878 (1994)
2. Barrett, S.: Environment & Statecraft: the Strategy of Environmental Treaty-Making. Oxford University Press (2005)
3. Finus, M.: Game Theoretic Research on the Design of International Environmental Agreements: Insights, Critical Remarks, and Future Challenges. International Review of Environmental and Resource Economics 2, 29–67 (2008)
4. Foster, D., Young, P.: Stochastic evolutionary game dynamics. Theoretical Population Biology 38(2), 219–232 (1990)
5. Fudenberg, D., Maskin, E.: The folk theorem in repeated games with discounting or with incomplete information. Econometrica 54, 533–554 (1986)
6. Kolstad, C.D.: Systematic Uncertainty in Self-enforcing International Environmental Agreements. Journal of Environmental Economics and Management 53, 68–79 (2007)
7. Nash, J.: The Bargaining Problem. Econometrica 18, 155–162 (1950)
8. Rubio, S., Ulph, A.: Leadership and Self-Enforcing International Environmental Agreements with Non-Negative Emissions. Unpublished manuscript, University of Valencia (2003)
9. Young, P.: The evolution of conventions. Econometrica 61, 57–84 (1993)
10. Young, P.: Individual strategy and social structure, pp. 113–129. Princeton University Press (1998)

Learn-and-Optimize: A Parameter Tuning Framework for Evolutionary AI Planning

Mátyás Brendel and Marc Schoenauer

Projet TAO, INRIA Saclay & LRI
matthias.brendel@lri.fr, marc.schoenauer@inria.fr

Abstract. Learn-and-Optimize (LaO) is a generic surrogate based method for parameter tuning combining learning and optimization. In this paper LaO is used to tune Divide-and-Evolve (DaE), an Evolutionary Algorithm for AI Planning. The LaO framework makes it possible to learn the relation between some features describing a given instance and the optimal parameters for this instance, thus it enables to extrapolate this relation to unknown instances in the same domain. Moreover, the learned knowledge is used as a surrogate-model to accelerate the search for the optimal parameters. The proposed implementation of LaO uses an Artificial Neural Network for learning the mapping between features and optimal parameters, and the Covariance Matrix Adaptation Evolution Strategy for optimization. Results demonstrate that LaO is capable of improving the quality of the DaE results even with only a few iterations. The main limitation of the DaE case-study is the limited amount of meaningful features that are available to describe the instances. However, the learned model reaches almost the same performance on the test instances, which means that it is capable of generalization.

1 Introduction

Parameter tuning is basically a general optimization problem applied off-line to find the best parameters for complex algorithms, for example for Evolutionary Algorithms (EAs). Whereas the efficiency of EAs has been demonstrated on several application domains [25,14], they usually need computationally expensive parameter tuning. Being a general optimization problem, there are as many parameter tuning algorithms as optimization techniques. However, several specialized methods have been proposed, and the most prominent ones today are Racing [4], REVAC [16], SPO [2], and ParamILS [10]. All these approaches face the same crucial generalization issue: can a parameter set that has been optimized for a given problem be successfully used for another one? The answer of course depends on the similarity of both problems.

J.-K. Hao et al. (Eds.): EA 2011, LNCS 7401, pp. 145–155, 2012.
© Springer-Verlag Berlin Heidelberg 2012

However, until now, in AI Planning, sufficiently accurate features have not been specified that would allow to describe the problem, no design of a general learning framework has been proposed, and no general experiments have been carried out. This paper makes a step toward a framework for parameter tuning applied generally to AI Planning and proposes a preliminary set of features. The Learn-and-Optimize (LaO) framework consists of the combination of optimizing and learning, i.e., finding the mapping between features and best parameters. Furthermore, the results of learning will already be useful to further the optimization phases, using the learned model similarly, but also in a different way as in standard surrogate-model based techniques (see e.g., [1] for a Gaussian-process-based approach).

In this paper, the target optimization technique is an Evolutionary Algorithm (EA), more precisely the evolutionary AI planner called Divide-and-Evolve (DaE). However, DaE will be here considered as a black-box algorithm, without any modification for the purpose of this work compared to its original version described in [13].

The paper is organized as follows: AI Planning Problems and the Divide-and-Evolve algorithm are briefly introduced in section 2. Section 3 introduces the original, top level parameter tuning method, Learn-and-Optimize. The case study presented in Section 4 applies LaO to DaE, following the rules of the International Planning Competition 2011 – Learning Track. Finally, conclusions are drawn and further directions of research are proposed in Section 5.

2   AI Planning and Divide-and-Evolve

An Artificial Intelligence (AI) planning problem is defined by the triplet of an initial state, a goal state, and a set of possible actions. An action modifies the current state and can only be applied if certain conditions are met. A solution plan to a planning problem is an ordered list of actions, whose execution from the initial state achieves the goal state.

Domain-independent planners rely on the Planning Domain Definition Language PDDL2.1 [6]. The domain file specifies object types and predicates, which define possible states, and actions, which define possible state changes. The instance scenario declares the actual objects of interest, sets the initial state and provides a description of the goal. A solution plan to a planning problem is a consistent schedule of grounded actions whose execution in the initial state leads to a state that contains the goal state, i.e., where all atoms of the problem goal are true. A planning problem defined on domain D with initial state I and goal G will be denoted in the following as $\mathcal{P}_D(I, G)$.

Early approaches to AI Planning using Evolutionary Algorithms directly handled possible solutions, i.e. possible plans: an individual is an ordered sequence of actions see [20,15,22,23,5]. However, as it is often the case in Evolutionary Combinatorial optimization, those direct encoding approaches have limited performance in comparison to the traditional AI planning approaches. Furthermore, hybridization with classical methods has been the way to success in many combinatorial domains, as witnessed by the fruitful emerging domain of memetic algorithms [9]. Along those lines, though relying on an original "memetization" principle, a novel hybridization of Evolutionary Algorithms (EAs) with AI Planning, termed Divide-and-Evolve (DaE) has been proposed [18,19]. For a complete formal description, see [12].

The basic idea of DaE in order to solve a planning task $\mathcal{P}_D(I, G)$ is to find a sequence of states $S_1, \ldots, S_n$, and to use some embedded planner to solve the series of planning problems $\mathcal{P}_D(S_k, S_{k+1})$, for $k \in [0, n]$ (with the convention that $S_0 = I$ and $S_{n+1} = G$). The generation and optimization of the sequence of states $(S_i)_{i \in [1,n]}$ is driven by an evolutionary algorithm. The fitness (makespan or total cost) of a list of partial states $S_1, \ldots, S_n$ is computed by repeatedly calling the external 'embedded' planner to solve the sequence of problems $\mathcal{P}_D(S_k, S_{k+1})$, $\{k = 0, \ldots, n\}$. The concatenation of the corresponding plans (possibly with some compression step) is a solution of the initial problem. Any existing planner can be used as embedded planner, but since guarantee of optimality at all calls is not mandatory in order for DaE to obtain good quality results [12], a suboptimal, but fast planner is used: YAHSP [21] is a lookahead strategy planning system for sub-optimal planning which uses the actions in the relaxed plan to compute reachable states in order to speed up the search process.

One-point crossover is used, adapted to variable-length representation in that both crossover points are independently chosen, uniformly in both parents. Four different mutation operators have been designed, and once an individual has been chosen for mutation (according to a population-level mutation rate), the choice of which mutation to apply is made according to user-defined relative weights. Because an individual is a variable length list of states, and a state is a variable length list of atoms, the mutation operator can act at both levels: at the individual level by adding (addState) or removing (delState) a state; or at the state level by adding (addAtom) or removing (delAtom) some atoms in the given state.

3   Learn-and-Optimize for Parameter Tuning

3.1   The General LaO Framework

As already mentioned, parameter tuning is actually a general global optimization problem, thus facing the routine issue of local optimality. But a further problem arises in parameter tuning, and this is the generality of the tuned parameters. Generalizing parameters learned on one instance to another instance might be problematic, because there are instances with very different complexity in the same domain. For example in [3] per-domain tuning was performed with the most difficult, largest instance, considered as a representative of the whole domain. However, it is clear from the results that these parameters were often suboptimal for the other instances. One workaround to this generalization issue is to relax the constraint of finding a single universally optimal parameterset, that certainly does not exist, and to focus on learning a complex relation between instances and optimal parameters. The proposed Learn-and-Optimize framework (LaO) aims at learning such a relation by adding learning to optimization. The underlying hypothesis is that there exists a relation between some features describing an instance and the optimal parameters for solving this instance which can be learned, and the goal of this work is to propose a general methodology to do so.

Suppose for now that we have n features and m parameters, and we are doing per-instance parameter tuning on instance $\mathcal{I}$. For the sake of simplicity and generality, both the fitness, the features and the parameters are considered as real values. Parameter tuning is the optimization (e.g., minimization) of the fitness function $f_\mathcal{I} : \mathbf{R}^m \to \mathbf{R}$, the expected value of the stochastic algorithm DaE executed with parameter $p \in \mathbf{R}^m$. The optimal parameter set is defined by $p_{opt} = argmin_p\{f_\mathcal{I}(p)\}$. For each instance $\mathcal{I}$, consider the set $F(\mathcal{I}) \in \mathbf{R}^n$ of the features describing this instance. Two relations have to be taken into account: each planning instance has features, and it has an optimal parameter-set. In order to be able to generalize, we have to get rid of the instance, and collapse both relations into one single relation between feature-space and parameter-space. For the sake of simplicity let us assume that there exists an unambiguous mapping from the feature space to the optimal parameter space.

$$p : \mathbf{R}^n \to \mathbf{R}^m, p(F) = p_{opt}. \tag{1}$$

The relation p between features and optimal parameters can be learned by any supervised learning method capable of representing, interpolating and extrapolating $\mathbf{R}^n \to \mathbf{R}^m$ mappings, provided sufficient data are available.

The idea of using some surrogate model in optimization is not new. Here, however, there are several instances to optimize, and only one model is available, that maps the feature-space into the parameter-space. A significant conceptual difference is that while in standard surrogate techniques the model is trained and evaluated for fitness, while in our approach we directly evaluate it for optimal parameter candidates.

Nevertheless, there is no question about how to use our model of p in optimization: one can always ask the model for hints about a given parameter-set. It seems reasonable that the stopping criterion of LaO is determined by the stopping criterion of the optimizer algorithm. After exiting one can also do a re-training of the learner with the best parameters found.

3.2   An Implementation of LaO

A simple multilayer Feed-Forward Artificial Neural Network (ANN) trained with standard backpropagation was chosen here for the learning of the features-to-parameters mapping, though any other supervised-learning algorithm could have been used. The implicit hypothesis is that the relation p is not very complex, which means that a simple ANN may be used. In this work, one mapping is trained for each domain. Training a single domain-independent ANN is left for future work. The other decision for LaO implementation is the choice of the optimizer used for parameter tuning. Because parameter optimization will be done successively for several instances, the simple yet robust (1+1)-Covariance Matrix Adaptation Evolution Strategy [8], in short (1+1)-CMA-ES, was chosen, and used with its robust own default parameters, as advocated in [3].

One original component, though, was added to some direct approach to parameter tuning: gene-transfer between instances. There will be one (1+1)-CMA-ES running for each instance, because using larger population sizes for a single instance would be far too costly. However, the (1+1)-CMA-ES algorithms running on all training instances form a population of individuals. The idea of gene-transfer is to use something like a crossover between the individuals of this population. Of course, the optimal parameter sets for the different instances are different; However, good 'chromosomes' for one instance may at least help another instance. Thus it may be used as a hint in the optimization of that other instance. Therefore random gene-transfer was used in the present implementation of LaO, by calling the so-called *Genetransferer*. This is similar to the migration operator of the so called Island Model Genetic Algorithm [24], and the justification is similar: parallelism and separability. There are however considerable differences: in our case, When the Genetransferer is requested for a hint for one instance, it returns the so-far best parameter of a different instance chosen with uniform random distribution (preventing, of course, that the default parameters are tried twice). Another benefit of Genetransferer is that it may smoothen out the ambiguities between instances, by increasing the probability for instances with the same features to test the same parameters, and thus the possibility to find out that the same parameters are appropriate for the same features. Figure 1 shows the LAO framework with the described implementations.

One additional technical difficulty arose with CMA-ES: each parameter is here restricted to an interval. This seems reasonable and makes the global algorithm more stable. Hence the parameters of the optimizer are actually normalized

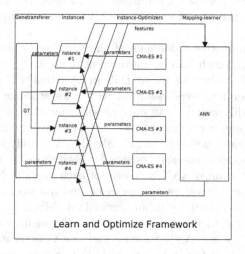

Fig. 1. Flowchart of the Lao framework, displaying only 4 instances

linearly onto the [0,1] interval. It is hence possible to apply a simple version of the box constraint handling technique described in [7], with a penalty term simply defined by $||p^{feas} - p||$, where p^{feas} is the closest value in the box. Moreover, only p^{feas} was recorded as a feasible solution , and later passed to the ANN. Note that the GeneTransferer and the ANN itself cannot return hints outside of the box. In order to not to compromise too much CMA-ES, several iterations of this were carried out for one hint of the ANN and one Genetransferer.

The implementation of LaO algorithm uses the Shark library [11] for CMA-ES and the FANN library for ANN [17]. To evaluate each parameter-setting with each instance, a cluster was used, that has approximately 60 nodes, most of them with 4 cores, some with 8. Because of the heterogeneity of the hardware architecture used here, it is not possible to rely on accurate predicted running times. Therefore, for each evaluation, the number of YAHSP evaluations in DaE is fixed. Moreover, since DaE is not deterministic, 11 independent runs were carried out for each DaE experiment with a given parameter-set, and the fitness of this parameter set was taken to be the median.

4   Results

In the Planning and Learning Part of IPC2011 (IPC), 5 sample domains were pre-published, with a corresponding problem-generator for each domain: Ferry, Freecell, Grid, Mprime, and Sokoban. Ferry and Sokoban were excluded from this study since there were not enough number of instances to learn any mapping. For each of the remaining 3 domains, approximately 100 instances were generated, with the published generators and distribution (ranges) of generator-parameters 100 instances per domain seemed to be appropriate for a running time of 2-3 weeks. The competition track description fixes running time as 15 minutes. However, many instances were never solved within 15 minutes, and those instances were dropped from the rest of experiment. The remaining instances were used for training.

The real IPC competition domains of the same track were released later. These domains were much harder, meaning that most of the official train instances could not be solved at all by DaE in 15 minutes. Therefore the published instance-generators were used, but with a lower range of the generator-parameters. Even this way, we can only present one domain: Parking. For the other domains, the number of training instances, or the iterations of LaO carried out until the deadline is not sufficient to take the case-study seriously.

Table 1 presents the train- and testsets for each domain. The Mean Square Error (MSE) of the trained ANN is shown for each domain. Note that because the fitness takes only few values, there can be multiple optimal parameter sets for the same instance, resulting in an unavoidable MSE. So we do not expect this error to converge to 0. One iteration of LaO amounts to 5 iterations of CMA-ES, followed by one ANN training and one Genetransferer. Due to the time constraints, only a few iterations of LaO were run. For example the 10 iterations in domain Grid amounts to 500 CMA-ES calls in total.

Table 1. Description of the domains

Domain Name	# of training instances	# of test instances
Freecell	108	230
Grid	55	124
Mprime	64	152
Parking	101	129

Table 2. DaE parameters that are controlled by LaO

Name	Min	Max	Default
Probability of crossover	0.0	1	0.8
Probability of mutation	0.0	1	0.2
Rate of mutation add station	0	10	1
Rate of mutation delete station	0	10	3
Rate of mutation add atom	0	10	1
Rate of mutation delete atom	0	10	1
Mean average for mutations	0.0	1	0.8
Time interval radius	0	10	2
Maximum number of stations	5	50	20
Maximum number of nodes	100	100 000	10 000
Population size	10	300	100
Number of offsprings	100	2 000	700

The controlled parameters of DaE are described in table 2. For a detailed description of these parameters, see [3]. The feature-set consists of 12 features which are presented in table 3. The first 5 features are computed from the domain file, after the initial grounding of YAHSP: number of fluents, goals, predicates, objects and types. One further feature we think could even be more important is called mutex-density, which is the number of mutexes divided by the number of all fluent-pairs. We also kept 6 less important features: number of lines, words and byte-count - obtained by the Linux-command "wc" - of the instance and the domain file. These features were kept only for historical reasons: they were used in the beginning as some "dummy" features. Note that some features take only one values, they however had meaning when training an inter-domain model.

The ANN had 3 fully connected layers, the layers had all 12 neurons, corresponding to the number or parameters and features, respectively. Standard back-propagation algorithm was used for learning (the default in FANN). In one iteration of LaO, the ANN was only trained for 50 iterations (aka epochs) without reseting the weights, in order to i- avoid over-training, and ii- making a gradual transition from the previous best parameter-set to the new best one, and eventually try some intermediate values. Hence, over the 10 iterations of LaO, 500 iterations (epochs) of the ANN were carried out in total. However, note that the best parameters were trained with much fewer iterations, depending on the time of their discovery. In the worst case, if the best parameter was found in the

Table 3. Minimum, mean and maximum values are given for the Freecell domain

Name	minimum	mean	maximum
# initial fluents	28	31.17	34
# goals	2	2	2
# predicates	7	7	7
# objects	32	35.17	38
# types	3	3	3
mutexdensity	0.14	0.15	0.17
# lines domain	198	198	198
# words domain	640	640	640
# bytes domain	5729	5729	5729
# lines instance	139	153.33	166
# words instance	379	427.42	469
# bytes instance	2017	2265.75	2479

last iteration of LaO, it was trained for only 50 epochs and not used anymore. This explains why retraining is needed in the end.

LaO has been running for several weeks on a cluster. But this cluster was not dedicated to our experiments, i.e. only a small number of 4 or 8-core processors were available for each domain on average. After stopping LaO, retraining was made with 300 ANN epochs with the best data, because the ANN's saved directly from LaO may be under-trained. The MSE error in retraining of the ANN did not decrease using more epochs, which indicates that 300 iterations are enough at least for this amount of data and for this size of the ANN. Tests with 1000 iterations did not produce better results and neither did the training of the ANN uniquely, i.e. only with the first found best parameters.

Since testing was also carried out on the cluster, the termination criterion for testing was also the number of evaluations for each instance. For evaluation the quality-improvement the quality-ratio metric defined in IPC competitions was used. The baseline qualities come from the default parameter-setting obtained by tuning for some representative domains with global racing in previous work see [3]. The ratio of the fitness value for the default parameter and the tuned parameter was computed and average was taken over the instances in the train or test-set.

$$Q = \frac{Fitness_{baseline}}{Fitness_{tuned}} \qquad (2)$$

Table 4 presents several quality-improvement ratios. Label "in LaO" means that the best found parameter is compared to the default. By definition, this ratio can never be less than 1. This improvement indicated by high quality-ratio is already useful if the very same instances used in training have to be optimized. Quality-improvement ratios for the retrained ANN on both the training-set and the test-set are also presented. In these later cases, numbers less then 1 are possible (the parameters resulting from the retrained ANN can have worse results than the ones given by the original ANN), but were rare.

Table 4. Results by domains (only the actually usable training instances are shown). ANN-error is given as MSE, as returned by FANN. The quality-improvement ratio in Lao is that of the best parameter-set found by LaO.

Domain Name	# of iterations	ANN error	quality-ratio in LaO	quality-ratio ANN on train	quality-ratio ANN on test
Freecell	16	0.1	1.09	1.05	1.04
Grid	10	0.09	1.09	1.05	1.03
Mprime	8	0.08	1.11	1.05	1.04
Parking	11	0.12	1.49	1.41	1.14

The first 3 domains in Table 4 have similar results: some quality-gain in training was consistently achieved, but the transfer of this improvement to the ANN-model was only partial. The phenomenon can appear because of the unambiguity of the mapping, or because the ANN is not complex enough for the mapping, or, and most probably, because the feature-set is not representative enough. On the other hand, the ANN model generalizes excellently to the independent test-set, at least for the first 3 domains. Quality-improvement ratios dropped only by 0.01, i.e. the knowledge incorporated in the ANN was transferable to the test cases and usable almost to the same extent than for the train set. The size of the training set seems not to be so crucial. For example for Freecell all the instances (108 out of 108 generated) could be used, because they were not so hard. On the other hand, only few Grid instances (55 out of 107 generated) could be used. However, both performed well. The explanation for this may be that both the 32 and 108 instances covered well the whole range of solvable instances. The situation is somewhat different for the last domain: Parking. Here we have a high quality-gain in training (almost 50%), even much of this could be learned by the ANN, however, here we have a huge drop for the test-set. The reason for this is that when using the ANN as an extrapolation, there is a considerable number of instances which get unsolvable. Still, we get some overall gain in average, in fact we get the highest gain for this difficult domain. The main issue for such hard domains will be to avoid much more effectively unfeasible parameters when extrapolating the ANN-model for unknown instances.

5   Conclusions and Future Work

The LaO method presented in this paper is a surrogate-model based combined learner and optimizer for parameter tuning. LaO was demonstrated to be capable of improving the quality of the DaE algorithm over tuning with global racing consistently, even though it was run only for a few iterations. Ongoing work is concerned with running LaO for an appropriate number of iterations. A clearly visible result is also that some of this quality-improvement can be incorporated into an ANN-model, which is also able to generalize excellently to an independent test-set.

The most important experiment to carry out in the future is simply to test the algorithm with more iterations and on more domains – and this will take several months of CPU even using a large cluster. Since LaO is only a framework, as indicated other kind of learning methods, and other kind of optimization techniques may be incorporated. If an ANN is used, the optimal structure has to be determined, or a more sophisticated solution is to apply one of the so-called Growing Neural Network architectures. Also the benefit of gene-transfer and/or crossover should be investigated further. Gene-transfer shall be improved so that chromosomes are transfered deterministically, measuring the similarity of instances by the similarity of their features. Present results indicate that the current feature set is too small and should be extended for better results. Also a more effective mechanism to avoid unfeasible parameters in the ANN-model has to be developed, especially for hard domains and instances.

Acknowledgements. This work is funded through French ANR project DESCARWIN ANR-09-COSI-002.

References

1. Bardenet, R., Kégl, B.: Surrogating the surrogate: accelerating gaussian-process-based global optimization with a mixture cross-entropy algorithm. In: Proceedings of the 27th International Conference on Machine Learning, ICML 2010 (2010)
2. Bartz-Beielstein, T., Lasarczyk, C., Preuss, M.: Sequential parameter optimization. In: McKay, B. (ed.) Proc. CEC 2005, pp. 773–780. IEEE Press (2005)
3. Bibai, J., Savéant, P., Schoenauer, M., Vidal, V.: On the generality of parameter tuning in evolutionary planning. In: Branke, J., et al. (eds.) Genetic and Evolutionary Computation Conference (GECCO), pp. 241–248. ACM Press (July 2010)
4. Birattari, M., Stützle, T., Paquete, L., Varrentrapp, K.: A Racing Algorithm for Configuring Metaheuristics. In: GECCO 2002, pp. 11–18. Morgan Kaufmann (2002)
5. Brié, A.H., Morignot, P.: Genetic Planning Using Variable Length Chromosomes. In: Proc. ICAPS (2005)
6. Fox, M., Long, D.: PDDL2.1: An Extension to PDDL for Expressing Temporal Planning Domains. JAIR 20, 61–124 (2003)
7. Hansen, N., Niederberger, S., Guzzella, L., Koumoutsakos, P.: A method for handling uncertainty in evolutionary optimization with an application to feedback control of combustion. IEEE Transactions on Evolutionary Computation 13(1), 180–197 (2009)
8. Hansen, N., Ostermeier, A.: Completely derandomized self-adaptation in evolution strategies. Evolutionary Computation 9(2), 159–195 (2001)
9. Hart, W., Krasnogor, N., Smith, J. (eds.): Recent Advances in Memetic Algorithms. STUDFUZZ, vol. 166. Springer, Heidelberg (2005)
10. Hutter, F., Hoos, H.H., Leyton-Brown, K., Stützle, T.: ParamILS: an automatic algorithm configuration framework. Journal of Artificial Intelligence Research 36, 267–306 (2009)
11. Igel, C., Glasmachers, T., Heidrich-Meisner, V.: Shark. Journal of Machine Learning Research 9, 993–996 (2008)
12. Bibai, J., Savéant, P., Schoenauer, M., Vidal, V.: An evolutionary metaheuristic based on state decomposition for domain-independent satisficing planning. In: ICAPS 2010, pp. 18–25. AAAI Press (2010)

13. Bibai, J., Savéant, P., Schoenauer, M., Vidal, V.: On the Benefit of Sub-optimality within the Divide-and-Evolve Scheme. In: Cowling, P., Merz, P. (eds.) EvoCOP 2010. LNCS, vol. 6022, pp. 23–34. Springer, Heidelberg (2010)
14. Lobo, F., Lima, C., Michalewicz, Z. (eds.): Parameter Setting in Evolutionary Algorithms. Springer, Berlin (2007)
15. Muslea, I.: SINERGY: A Linear Planner Based on Genetic Programming. In: Steel, S. (ed.) ECP 1997. LNCS, vol. 1348, pp. 312–324. Springer, Heidelberg (1997)
16. Nannen, V., Smit, S.K., Eiben, A.E.: Costs and Benefits of Tuning Parameters of Evolutionary Algorithms. In: Rudolph, G., Jansen, T., Lucas, S., Poloni, C., Beume, N. (eds.) PPSN 2008. LNCS, vol. 5199, pp. 528–538. Springer, Heidelberg (2008)
17. Nissen, N.: Implementation of a Fast Artificial Neural Network Library (FANN). Technical report, Department of Computer Science University of Copenhagen, DIKU (2003)
18. Schoenauer, M., Savéant, P., Vidal, V.: Divide-and-Evolve: A New Memetic Scheme for Domain-Independent Temporal Planning. In: Gottlieb, J., Raidl, G.R. (eds.) EvoCOP 2006. LNCS, vol. 3906, pp. 247–260. Springer, Heidelberg (2006)
19. Schoenauer, M., Savéant, P., Vidal, V.: Divide-and-Evolve: a Sequential Hybridization Strategy using Evolutionary Algorithms. In: Michalewicz, Z., Siarry, P. (eds.) Advances in Metaheuristics for Hard Optimization, pp. 179–198. Springer (2007)
20. Spector, L.: Genetic Programming and AI Planning Systems. In: Proc. AAAI 1994, pp. 1329–1334. AAAI/MIT Press (1994)
21. Vidal, V.: A lookahead strategy for heuristic search planning. In: Proceedings of the 14th International Conference on Automated Planning and Scheduling (ICAPS 2004), Whistler, BC, Canada, pp. 150–159. AAAI Press (June 2004)
22. Westerberg, C.H., Levine, J.: "GenPlan": Combining Genetic Programming and Planning. In: Garagnani, M. (ed.) 19th PLANSIG Workshop (2000)
23. Westerberg, C.H., Levine, J.: Investigation of Different Seeding Strategies in a Genetic Planner. In: Boers, E.J.W., Gottlieb, J., Lanzi, P.L., Smith, R.E., Cagnoni, S., Hart, E., Raidl, G.R., Tijink, H. (eds.) EvoWorkshops 2001. LNCS, vol. 2037, pp. 505–514. Springer, Heidelberg (2001)
24. Whitley, D., Rana, S., Heckendorn, R.B.: The island model genetic algorithm: On separability, population size and convergence. Journal of Computing and Information Technology 7, 33–47 (1998)
25. Yu, T., Davis, L., Baydar, C., Roy, R. (eds.): Evolutionary Computation in Practice. SCI, vol. 88. Springer, Heidelberg (2008)

A Model Based on Biological Invasions
for Island Evolutionary Algorithms

Ivanoe De Falco[1], Antonio Della Cioppa[2], Domenico Maisto[1],
Umberto Scafuri[1], and Ernesto Tarantino[1]

[1] ICAR–CNR, Via P. Castellino 111, 80131 Naples, Italy
{ivanoe.defalco,domenico.maisto,umberto.scafuri,
ernesto.tarantino}@na.icar.cnr.it
[2] DIEII, University of Salerno, Via Ponte don Melillo 1, 84084 Fisciano (SA), Italy
adellacioppa@unisa.it

Abstract. Migration strategy plays an important role in designing
effective distributed evolutionary algorithms. Here, a novel migration
model inspired to the phenomenon known as biological invasion is
adopted. The migration strategy is implemented through a multistage
process involving large invading subpopulations and their competition
with native individuals. In this work such a general approach is used
within an island parallel model adopting Differential Evolution as the
local algorithm. The resulting distributed algorithm is evaluated on a
set of well known test functions and its effectiveness is compared against
that of a classical distributed Differential Evolution.

Keywords: massive migration, biological invasion, distributed EAs.

1 Introduction

Evolutionary Algorithms (EAs) [1] have shown to be very effective to solve
complex search and optimization problems in different fields. Their main
drawback is related to the convergence speed. A popular way to enhance EA
performance is to implement parallel versions in which the individuals are
distributed over several subpopulations (*islands*) which evolve independently and
interact by a migration operator used to exchange individuals among them. This
distributed framework is based on the classical coarse-grained approach to EAs
[2]. It consists in a locally-linked strategy, known as *stepping stone-model* [3], in
which each EA instance is connected to a number of other instances equal to the
degree of connectivity of the topology beneath. Thus, each subpopulation is kept
relatively 'isolated' from all the other ones in the sense that it can be implicitly
influenced by them only through communications with its own neighbors.

The selection function, deciding what individuals migrate, and the
replacement function, indicating the individuals to be substituted by the
immigrants, play a key role for the effectiveness of the model. Different strategies
can be devised: the migrants can be selected either according to better fitness
values or randomly, and they might replace the worst individuals or substitute

J.-K. Hao et al. (Eds.): EA 2011, LNCS 7401, pp. 157–168, 2012.

them only if better, or they might replace any individual (apart from the best ones) in the neighboring subpopulations. The exchange of the solutions is determined by the *migration rate* defining the number of individuals that are sent to (received from) other islands, a *replacement function* implementing the substitution policy and by the *migration interval* indicating the exchange frequency among neighboring subpopulations. An unsuitable number of migrants can be disruptive when using swap, substitution or copy as replacement function [4,5]. Finally, the island topology determines the number of the neighboring subpopulations and its diameter gives an indication of the speed at which information spreads through them [6].

Due to interaction among subpopulations, island EAs have a dynamic behavior totally different from that of their sequential versions. They are likely to explore a search space more evenly and may face population stagnation thanks to their ability to maintain population diversity [6]. Another advantage is that generally these distributed versions allow speedup in computation and better solution quality [7] in comparison with a single EA evolution.

The idea underlying this paper is to device a novel migration model inspired to the biological phenomenon known as *biological invasion*, consisting in the introduction, the establishment and the spread of non–native organisms in an environment [8]. Although this process has inherently ecological impacts, it has also some evolutionary consequences. Essentially, when organisms are introduced to a new habitat they may experience new selective pressures due to biotic (i.e., indigenous species) and abiotic (non–living chemical and physical environmental components) factors, and simultaneously act as novel selective agents on native individuals in the invaded ecosystem. Such conditions are extremely favorable for a rapid evolution of both the invaders and the natives they interact with in the new environment [9]. This frequently induces a quick adaptation of both native and alien organisms to prevent their extinction [9,8,10]. According to the literature produced in this field, biological invasions can be divided into different stages [11,12]. Firstly, a *propagule*, i.e. a group of individuals involved in the invasion, is formed in its native habitat. Secondly, that propagule is transported to a new range where forms a *founding subpopulation*. Finally, the establishment and the spread of the propagule in the new habitat take place by means of a competition process with native individuals. Each stage of this process acts as a filter, passing only individuals that own the characteristics needed to survive to novel selection pressures. The impact of a biological invasion is assessed through a composite measure named as *propagule pressure*. It relies on two key components: the propagule size (the number of individuals composing a propagule) and the propagule number (the rate at which propagules are introduced per unit of time). Thus, propagule pressure is the distribution of propagule sizes and the frequency in which propagules arrive [13]. Both theoretical and empirical studies suggest that the propagule pressure is often positively correlated with the success of invasion process [14,12]. Genetic studies conducted recently have confirmed this hypothesis showing that large invading populations and multiple introductions are required for the process to succeed in each stage [14].

With the aim to exploit the above mentioned features of *biological invasion*, we have introduced into a generic island EA, instead of the canonical exchange process, a migration model inspired to the invasion process, structured in three distinct stages. With reference to each island, these stages are: the formation of the propagules in neighboring islands, the creation of a founding subpopulation and, finally, the generation of a new local subpopulation. In particular, the individuals to migrate from a subpopulation are those which are fitter than the current local average fitness. Then a founding subpopulation is formed by collecting the immigrants from all the neighbors and the native individuals. Finally, to simulate the establishment and the spread of the propagule and, hence, to generate a new local subpopulation, a competition replacement mechanism is adopted, i.e., the new subpopulation is created by choosing, through a fitness-proportionate selection, among all the individuals in the founding subpopulation. Such a novel general Invasion–based Model for distributed Evolutionary Algorithm (IM–dEA), can be conceived for use with any EA. In this paper, simulations have been carried out by choosing as evolutionary algorithm the Differential Evolution (DE) [15,16]. To assess the effectiveness of IM–dEA a comparison is carried out on a set of diverse benchmark problems with respect to a distributed version of the DE algorithm, *i.e.*, DDE [17].

Paper structure is as follows: Section 2 outlines the related works; Section 3 presents the details of the parallel framework with the novel migration scheme. In Section 4 the experimental findings are discussed, and behavior of IM–dEA is described. Finally, Section 5 contains conclusions and future works.

2   Related Works

Several island EA models differing in the selection function, the replacement function, the migration rate and the topology have been proposed in literature.

Cantú–Paz [18] investigates how the policy used to select migrants and individuals to replace affects the selection pressure in parallel EAs. The four possible combinations of random and fitness–based migration and replacement of existing individuals in the subpopulations are taken into account. The paper shows that the migration policy that chooses both the migrants and the replacements according to their fitness increases the selection pressure and may cause a significatively faster convergence.

In [5] different migration intervals and migration sizes are investigated. In general, the authors suggest that the best performance is achieved with small migration sizes and moderate migration intervals. Their experiments indicate that the reasons for this migration policy are the necessity of maintaining diversity in the system to produce novel results and the need for enough information exchange among islands to combine partial results.

Considered that we have chosen DE for our investigation, some distributed DEs are reported. In [19] the migration mechanism as well as the algorithmic parameters are adaptively coordinated according to a criterion based on genotypical diversity. An adaptive DE is executed on each subpopulation for a fixed number of generations. Then a migration process, based on a random connection topology, is started: each individual in each subpopulation can be probabilistically swapped with a randomly selected individual in a randomly chosen subpopulation (including the one containing the initial individual).

Tasoulis et. al [20] propose a distributed DE characterized by a ring topology, a selection function that picks up the individuals with the best performance and a replacement function that substitutes random individuals of the neighboring subpopulation. A similar approach is shown in [21] with the same topology whereas the migration consists in the best individual replacing the oldest member of another subpopulation in the topological neighborhood.

In Apolloni et al. [22] a distributed version which modifies the one suggested in [20] is presented: the migration policy is based on a probabilistic criterion depending on five parameters. The individuals to migrate are randomly selected and the individuals arriving from other islands replace randomly chosen local individuals only if the former ones are fitter. The topology is a unidirectional ring in which the individuals are exchanged with the nearest neighbors.

In [23] a distributed DE algorithm with subpopulations grouped into two families is described. The first family has the role of exploring decision space and constituting an external evolutionary framework. In the second family, subpopulations reduce progressively their size, in order to increase highly exploitation and to quickly detect solutions with a high performance. Then the solutions generated by the second family migrate to the first family.

Weber et al. [24] propose a distributed DE that employs a novel self–adaptive scheme. Subpopulations are disposed in a ring topology. Each subpopulation has its own scale factor value. With a probabilistic criterion, the individual with the best performance migrates to the neighboring island and replaces a pseudo-randomly selected individual of the target subpopulation. The target subpopulation inherits also the scale factor if it is promising at the current stage of the evolution. A perturbation mechanism is introduced to improve the exploration ability of the algorithm.

Ishimizu and Tagawa [25] present a structured DE approach still based on the island model. Different network topologies are taken into account. The migration takes place every fixed number of generations and the exchange involves only the best individual which migrates towards only one of the adjacent subpopulations on the basis of the topological neighborhood and randomly replaces an individual, except for the best one, in the receiving subpopulation.

3   IM–dEA Scheme

The proposed algorithm distinguishes itself from other parallel evolutionary algorithms by both the number of individuals migrating from a node to another and by the way such individuals are used to update the receiving population. The philosophy underlying this novel migration scheme is inspired by biological invasion phenomena. As in the biological context as in the computational one, if the number of new individuals introduced in a population is commensurable with the size of receiving population, evolutionary pressure increases [18]. In reality, the arrival of alien individuals alters the balance of the new habitat insofar as they compete with internal individuals to survive.

Our model makes use of a locally connected topology where each node, i.e., each processor, hosts an instance of an EA, and is connected to other $|\mathscr{C}(p)|$ nodes, where $\mathscr{C}(p)$ is the set of p neighbors of the node.

Execution of the algorithm starts by means of an initialization process. Let us suppose there are $\mathscr{N}$ nodes in the chosen topology; $\mathscr{N}$ subpopulations $\{P_1, \ldots, P_p, \ldots, P_{\mathscr{N}}\}$, each one composed of n individuals, are randomly sampled and each of them is assigned to a different node.

At each generation, each subpopulation P_p performs a sequential EA. As a consequence, each subpopulation is updated from a generation to the next one by means of the steps typical of the EA chosen. This scheme is repeated for each individual and for a number of generations $\mathscr{T}$, named as *invasion time*. This corresponds to the migration interval usually defined for distributed EAs. Every $\mathscr{T}$ generations, neighboring subpopulations exchange individuals. The set of individuals each subpopulation P_p sends to its neighbors is called *propagule* and is indicated as $\mathscr{M}_{P_p}$. $\mathscr{M}_{P_p}$ is determined by collecting the individuals of P_p which are fitter than its current average fitness:

$$\mathscr{M}_{P_p} = \left\{ x_i^p \in P_p \;\middle|\; f(x_i^p) \succ \frac{1}{n} \sum_{j=1}^{n} f(x_j^p) \right\} \tag{1}$$

where $f(x_i^p)$ is the fitness associated to the individual x_i^p and "$\succ$" is a binary relation stating that the left member is fitter than the right member. All of the chosen individuals are sent to all of the neighboring subpopulations $P_{\tilde{p}}$, with $\tilde{p} \in \mathscr{C}(p)$, so, at each invasion time $\mathscr{T}$, each subpopulation receives a total number of $\sum_{\tilde{p}} |\mathscr{M}_{P_{\tilde{p}}}|$ elements where $|\mathscr{M}_{P_{\tilde{p}}}|$ is the cardinality of the set $\mathscr{M}_{P_{\tilde{p}}}$. By considering that each propagule originated in any of the neighboring nodes has a size averagely equal to circa $n/2$, the number of individuals introduced to each node is $|\mathscr{C}(p)| \cdot n/2$, approximately.

Afterwards, in each node, a larger subpopulation, formed by both native and exotic individuals, is constructed by adding this massive number of alien individuals to its own subpopulation. This new subpopulation is called *founding subpopulation* Π_p and is defined as:

$$\Pi_p \equiv \left\{ \bigcup_{\tilde{p} \in \mathscr{C}(p)} \mathscr{M}_{P_{\tilde{p}}} \right\} \cup P_p \tag{2}$$

with a size equal to $n + \sum_{\tilde{p}} |\mathscr{M}_{P_{\tilde{p}}}|$. Fundamentally, Π_p is an archive of potential solutions and is composed of both internal and external candidate solutions. Hence, Π_p constitutes a source of heterogeneity exploitable by the algorithm to improve its ability to search new evolutionary paths.

Algorithm 1. Pseudo-code of IM–dEA on a generic node

---

randomly generate an initial subpopulation P_p of n individuals
evaluate fitness of each individual
while halting conditions are not satisfied **do**
  for each generation **do**
    generate the new subpopulation through the classical actions of the chosen EA
    evaluate fitness of individuals in the new subpopulation
    if invasion time **then**
      create the propagule $\mathscr{M}_{P_p}$ for the current subpopulation
      send the propagule $\mathscr{M}_{P_p}$ to the neighboring subpopulations
      receive the propagules $\mathscr{M}_{P_{\tilde{p}}}$ from the neighboring subpopulations
      construct the founding subpopulation Π_p
      copy the best individual of Π_p in P'_p
      select $n-1$ individuals in Π_p through stochastic universal sampling
      insert the $n-1$ selected individuals into the new population P'_p
    end if
  end for
end while

Subsequently, a new subpopulation P'_p is built up for the current node. Firstly, elitism is performed by copying the best individual of Π_p in P'_p. Secondly, stochastic universal sampling [26] (SUS) is applied on Π_p and $n-1$ individuals selected in this way are inserted into P'_p. SUS permits to choose without bias potentially useful solutions from Π_p via a fitness-proportionate selection. This mechanism, interpretable as an additional reproduction induced by introduction of elements arriving from neighboring nodes, enhances both the exploitation and the exploration of the original algorithm and guarantees their trade-off. The described procedure, whose pseudo-code is represented in Algorithm 1, is repeated on each node until the halting conditions are satisfied.

At this point, it is worth stressing some remarkable differences between the particular 'migration' mechanisms characterizing IM–dEA and the other common distributed EAs based on the stepping stone-model. Usually, this kind of distributed EAs exchanges individuals, selected by some specific characteristic, between the islands with a ratio equal to one: each immigrated individual substitutes an individual in the receiving subpopulation. Besides, such a substitution mostly occurs by either a swap or a copy-and-replacement.

IM–dEA acts in a very different way. Here, exchange is performed in three distinct stages: 1) the formation of the propagules in neighboring islands according to equation (1); 2) the creation of a founding subpopulation as specified in equation (2) with a size much larger than that of the local subpopulation; 3) the generation of a new local subpopulation through the application of an elitism mechanism and SUS on the founding subpopulation. Each stage exerts a selective pressure not traceable in other distributed evolutionary algorithms where substitutions generally take place without any actual competition among individuals.

Ultimately, in our opinion, the model of exchange of the solutions among subpopulations, inspired to biological invasion phenomenon and presented in this work might be considered as an extra evolutionary mechanism with respect to the main one implemented by the specific evolutionary algorithm adopted.

4   Experiments

To investigate the effectiveness of IM–dEA scheme, we have decided to compare its behavior against that of DDE on a set of 12 benchmark functions. Namely, the 500–sized versions of Ackley, Griewangk, Michalewicz, Rastrigin, Rosenbrock and Schwefel functions, as well as those of their rotated versions, have been taken into account [23].

DDE consists of a set of classical DE schemes, running in parallel, assigned to different processing elements arranged in a torus topology [17], in which each generic DE instance has four neighboring communicating subpopulations, and migration rate and migration interval have been set equal to 1 and 5, respectively. The individual sent is the best one and it randomly replaces an individual in the neighboring subpopulation, except for the local current best one. The mutation mechanism used in DDE is $DE/rand/1/bin$ [15,16].

To carry out the comparison, we have implemented an IM–dEA containing instances of DE, hence we have named it IM–dDE. Throughout the experiments, the values for the parameters *crossover ratio* (CR) and *scale factor* (F) have been set to 0.3 and 0.7, respectively, as suggested in [23], for both algorithms, and the DE operator in IM–dDE is the same of DDE. A 4×4 folded torus has been chosen as topology for the network of subpopulations, resulting in a total of 16 nodes. The total population size has been chosen as 200, which results in eight subpopulations with 12 individuals and eight containing 13. The number of generations has been set to 2, 500.

A first phase of our investigation has aimed at finding the best possible value for the migration interval $\mathscr{T}$ for any function, and this for both algorithms. We have considered a given range of possible values, i.e., 5, 10, 20, 30, 40, 50, 60, 70, 80, 90 and 100. For any such value 25 runs have been effected for each function and each algorithm, and the averages A_v of the best final fitness values over the 25 runs have been computed. Table 1 reports the best values of $\mathscr{T}$, together with the corresponding values of A_v.

A very important remark from Table 1 is that for any problem, apart from Rastrigin, Rosenbrock, and Schwefel rotated, the best $\mathscr{T}$ for IM–dDE is lower than that for DDE. This seems to represent algorithmically the correlation between biological invasion success and propagule pressure, namely, large invading subpopulations and frequent introductions lead to a faster adaptation of both native and alien individuals.

Examination of the results shows that in several cases the best value for $\mathscr{T}$ is obtained at the lowest tested migration interval. For the remaining problems, it happens that the results are better and better as the migration interval increases, and this holds true until a given value for $\mathscr{T}$ is reached; after this value, the performance worsens more and more as $\mathscr{T}$ further increases.

Table 1. Best migration interval and related average final value for each problem

Problem	IM–dDE $\mathcal{T}$	IM–dDE A_v	DDE $\mathcal{T}$	DDE A_v	FACPDE A_v
Ackley	5	$1.26 \cdot 10^{-2}$	10	$6.65 \cdot 10^{-2}$	$2.67 \cdot 10^{-2}$
Griewangk	5	$1.77 \cdot 10^{+0}$	10	$1.65 \cdot 10^{+1}$	$5.04 \cdot 10^{+2}$
Michalewicz	20	$-3.44 \cdot 10^{+2}$	30	$-3.21 \cdot 10^{+2}$	$-3.54 \cdot 10^{+2}$
Rastrigin	20	$1.46 \cdot 10^{+3}$	20	$1.89 \cdot 10^{3}$	$9.91 \cdot 10^{+2}$
Rosenbrock	10	$6.38 \cdot 10^{+2}$	10	$1.59 \cdot 10^{+3}$	$1.63 \cdot 10^{+3}$
Schwefel	20	$-1.52 \cdot 10^{+5}$	30	$-1.43 \cdot 10^{+5}$	$-1.54 \cdot 10^{+5}$
Ackley rotated	5	$1.74 \cdot 10^{-2}$	10	$1.66 \cdot 10^{-1}$	$6.92 \cdot 10^{-2}$
Griewangk rotated	5	$1.69 \cdot 10^{+0}$	10	$1.82 \cdot 10^{+1}$	$5.10 \cdot 10^{+2}$
Michalewicz rotated	30	$-7.67 \cdot 10^{+1}$	60	$-8.62 \cdot 10^{+1}$	$-1.92 \cdot 10^{+2}$
Rastrigin rotated	20	$1.66 \cdot 10^{+3}$	30	$2.51 \cdot 10^{+3}$	$1.10 \cdot 10^{+3}$
Rosenbrock rotated	5	$5.32 \cdot 10^{+2}$	10	$9.76 \cdot 10^{+2}$	$9.74 \cdot 10^{+2}$
Schwefel rotated	30	$-9.45 \cdot 10^{+4}$	30	$-9.26 \cdot 10^{+4}$	$-1.67 \cdot 10^{+5}$

From the table, it can be seen that IM–dDE performance seems superior to that of DDE on any problem, apart from Michalewicz rotated. Namely, on five problems IM–dDE reaches results better by one order of magnitude.

To ascertain whether or not those figures allow us to conclude the superiority of the approach here proposed from a statistical viewpoint, Wilcoxon test [27] has been performed for any function by taking into account the 25 best final values achieved by both algorithms. The test is performed by taking as null hypothesis the one that the two algorithms are equivalent, at the 0.05 significance level. Results of the test are shown in Table 2 where, for each function, the values of Pr (probability of observing equivalence by chance if the null hypothesis is true) and H (1 means acceptance of the null hypothesis, 0 rejection) are reported. The last column contains a '+' if IM–dDE is better than DDE, a '−' in the opposite case, and a '=' if the two algorithms are equivalent.

The results indicate that the null hypothesis can be rejected for all the functions but Michalewicz rotated and Schwefel rotated, for which the results of IM–dDE and DDE should be considered as equivalent. Table 2 also says that for the only function for which DDE seems to be slightly better than IM–dDE, i.e., Michalewicz rotated, there is actually a substantial equivalence.

Finally, in Table 1 we have also reported the average performance of FACPDE [24], one of the most performing distributed DE. As it is evident, IM–dDE is very competitive and, in many cases, exhibits better performance.

4.1   IM–dDE Behavior and Its Motivations

A first remark about the behavior of our algorithm can be made by looking at Fig. 1. The left panel shows the evolution of a typical run on the Rastrigin function for both algorithms with $\mathcal{T} = 30$. It is worth noting that the general dynamics in terms of evolution is similar for both the algorithms, the difference being that IM–dDE gets advantage over DDE in the early generations, and keeps

Table 2. Results of Wilcoxon test for all the faced functions

Problem	Pr	H	result
Ackley	$1.41 \cdot 10^{-9}$	1	+
Griewangk	$1.41 \cdot 10^{-9}$	1	+
Michalewicz	$1.41 \cdot 10^{-9}$	1	+
Rastrigin	$1.41 \cdot 10^{-9}$	1	+
Rosenbrock	$1.41 \cdot 10^{-9}$	1	+
Schwefel	$5.21 \cdot 10^{-9}$	1	+
Ackley rotated	$1.41 \cdot 10^{-9}$	1	+
Griewangk rotated	$1.41 \cdot 10^{-9}$	1	+
Michalewicz rotated	$7.12 \cdot 10^{-1}$	0	=
Rastrigin rotated	$1.41 \cdot 10^{-9}$	1	+
Rosenbrock rotated	$1.41 \cdot 10^{-9}$	1	+
Schwefel rotated	$5.60 \cdot 10^{-1}$	0	=

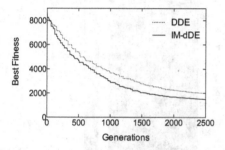 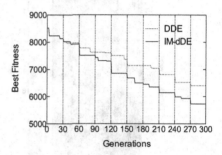

Fig. 1. Evolution of a typical run on the Rastrigin function. Left panel: the whole evolution. Right panel: zoom on the first 300 generations.

it during the whole evolution. It looks like IM–dDE somehow 'anticipates' DDE. The first part of this figure evidences the rapid adaptation of both native and alien individuals already described above. The right panel, instead, displays the first 300 generations only: noticeable improvements in best fitness can be noted after almost every migration.

A second important feature of IM–dDE can be realized by taking a closer look at the behavior of our algorithm as provided in Fig. 2 that makes reference again to a typical run on the Rastrigin function for $\mathscr{T} = 30$. The figure reports the number of immigrants accepted in P_p' on node 0 (P_0') at each migration. It is worth noting that this number varies within a minimum of 5 and a maximum of 12, and is usually between 6 and 10. So, the number of accepted migrants is noticeably higher in IM–dDE than it is in DDE. Since the subpopulation under investigation has 12 individuals, this means that at generations $1,350$ and $1,620$ the local population is completely replaced by immigrants.

A third extremely important feature of IM–dDE resides in the genotypic diversity. Figure 3 shows the genotypic variance for each gene in P_0 (in horizontal) as a function of generations (in vertical) for a typical run of both

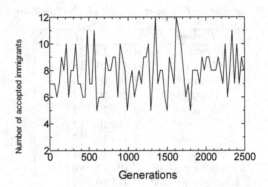

Fig. 2. The number of immigrants accepted in subpopulation P_0' at each migration

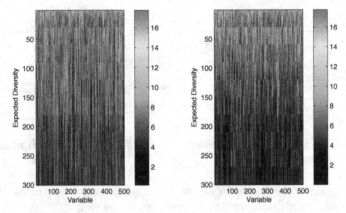

Fig. 3. Genotypic diversity in a subpopulation for DDE (left) and IM-dDE (right)

algorithms for Rastrigin function. Only the first 300 generations are reported. Migration takes place every 30 generations. A clear difference in genotypic variance before and after any migration can be appreciated in IM-dDE, since horizontal lines can be distinctly perceived at migrations. This says that the occurrence of each migration actually yields a net difference in genotypes, which results in a different regime in the subpopulation. The effects of migration are, instead, much less evident for DDE, for which a higher genotypic continuity is reported in figure, if we look at each gene, and horizontal lines are harder to detect at migrations.

As a further remark, we wish to say that we have also made some preliminary investigations about the use of DE operators other than the $DE/rand/1/bin$. They seem to confirm both the superiority of IM-dDE with respect to DDE in terms of higher solution quality and the above described IM-dDE behavior.

Finally, it should be noted that the highest computational cost of the migration model introduced by us is related to the sorting needed by SUS and is $O(n \log n)$. As regards the communication overhead, if Δt is the time required by MPI for sending just one individual, the time needed to send m individuals is between $m/2 \cdot \Delta t$ and $m \cdot \Delta t$, where $m \sim 6$ in our experiments.

5   Conclusions and Future Works

A new distributed scheme for EAs, IM-dDE, has been introduced, based on the concept of mimicking massive immigration phenomena. It has been used here to implement a distributed DE scheme. The resulting IM-dDE algorithm has been evaluated on a set of classical test functions. Its results, compared against those of DDE, have shown the effectiveness of the proposed approach.

Since the dynamics of IM-dDE is faster than that of DDE in achieving solutions with satisfactory quality, our algorithm could be profitably used to find suboptimal solutions to problems requiring quasi–realtime answers.

Future works will aim at carrying out a wider evaluation phase. This will be accomplished by performing sets of experiments with other DE operators, so as to ascertain IM-dDE performance with respect to that of DDE independently of the specific operator chosen, and to understand if some operators are, on average, better than the others for IM-dDE. Moreover we plan to make use of a wider set of test functions, and to take into account for comparison a larger number of other distributed versions of DE available in literature. Furthermore, dependence on the values of F and CR will be investigated.

Finally, since the scheme introduced is general and could be favorably adopted for any evolutionary algorithm, we aim at using it to implement distributed versions for other evolutionary algorithms as well.

References

1. Holland, J.: Adaptation in natural and artificial systems (1975)
2. Cantú-Paz, E.: A summary of research on parallel genetic algorithms. Technical Report 95007, University of Illinois, Urbana–Champaign, USA (1995)
3. Mühlenbein, H.: Evolution in time and space - the parallel genetic algorithm. In: Rawlins, G. (ed.) Foundation of Genetic Algorithms, pp. 316–337. Morgan Kaufmann, San Mateo (1992)
4. Grajdeanu, A.: Parallel models for evolutionary algorithms. Technical report, ECLab, Summer Lecture Series, George Mason University, 38 (2003)
5. Skolicki, K., De Jong, K.: The influence of migration sizes and intervals on island models. In: Proc. of the Conference of Genetic and Evolutionary Computation, pp. 1295–1302. Association for Computing Machinery, Inc. (ACM) (2005)
6. Tomassini, M.: Spatially structured evolutionary algorithms. Springer (2005)
7. Alba, E., Tomassini, M.: Parallelism and evolutionary algorithms. IEEE Trans. on Evolutionary Computation 6(5), 443–462 (2002)
8. Shigesada, N., Kawasaki, K.: Biological invasions: theory and practice. Oxford University Press, USA (1997)
9. Suarez, A., Tsutsui, N.: The evolutionary consequences of biological invasions. Molecular Ecology 17(1), 351–360 (2008)
10. Strayer, D., Eviner, V., Jeschke, J., Pace, M.: Understanding the long-term effects of species invasions. Trends in Ecology & Evolution 21(11), 645–651 (2006)

11. Catford, J., Jansson, R., Nilsson, C.: Reducing redundancy in invasion ecology by integrating hypotheses into a single theoretical framework. Diversity and Distributions 15(1), 22–40 (2009)
12. Kolar, C., Lodge, D.: Progress in invasion biology: predicting invaders. Trends in Ecology & Evolution 16(4), 199–204 (2001)
13. Simberloff, D.: The role of propagule pressure in biological invasions. Annual Review of Ecology, Evolution, and Systematics 40, 81–102 (2009)
14. Lockwood, J., Cassey, P., Blackburn, T.: The role of propagule pressure in explaining species invasions. Trends in Ecology & Evolution 20(5), 223–228 (2005)
15. Price, K., Storn, R.: Differential evolution. Dr. Dobb's Journal 22(4), 18–24 (1997)
16. Storn, R., Price, K.: Differential evolution – a simple and efficient heuristic for global optimization over continuous spaces. Journal of Global Optimization 11(4), 341–359 (1997)
17. De Falco, I., Della Cioppa, A., Maisto, D., Scafuri, U., Tarantino, E.: Satellite Image Registration by Distributed Differential Evolution. In: Giacobini, M. (ed.) EvoWorkshops 2007. LNCS, vol. 4448, pp. 251–260. Springer, Heidelberg (2007)
18. Cantú-Paz, E.: Migration policies, selection pressure, and parallel evolutionary algorithms. Journal of Heuristics 7(4), 311–334 (2001)
19. Zaharie, D.: Parameter adaptation in differential evolution by controlling the population diversity. In: Petcu, D., et al. (eds.) Proc. of the International Workshop on Symbolic and Numeric Algorithms for Scientific Computing, pp. 385–397 (2002)
20. Tasoulis, D., Pavlidis, N., Plagianakos, V., Vrahatis, M.: Parallel differential evolution. In: Proc. of the Congress on Evolutionary Computation, June 19-23, vol. 2, pp. 2023–2029 (2004)
21. Kozlov, K.N., Samsonov, A.M.: New migration scheme for parallel differential evolution. In: Proc. of the International Conference on Bioinformatics of Genome Regulation and Structure, pp. 141–144 (2006)
22. Apolloni, J., Leguizamón, G., García-Nieto, J., Alba, E.: Island based distributed differential evolution: an experimental study on hybrid testbeds. In: Proc. of the Eight International Conference on Hybrid Intelligent Systems, pp. 696–701. IEEE Press (2008)
23. Weber, M., Neri, F., Tirronen, V.: Distributed differential evolution with explorative-exploitative population families. Genetic Programming and Evolvable Machines 10(4), 343–371 (2009)
24. Weber, M., Neri, F., Tirronen, V.: Scale factor inheritance mechanism in distributed differential evolution. Soft Computing 14, 1187–1207 (2010)
25. Ishimizu, T., Tagawa, K.: A structured differential evolution for various network topologies. Int. Journal of Computers and Communications 4(1), 2–8 (2010)
26. Baker, J.: Reducing bias and inefficiency in the selection algorithm. In: Proc. of the Second International Conference on Genetic Algorithms and their Application, pp. 14–21. L. Erlbaum Associates Inc. (1987)
27. Wilcoxon, F.: Individual comparisons by ranking methods. Biometrics Bulletin 1(6), 80–83 (1945)

A Multi-objective Particle Swarm Optimizer Enhanced with a Differential Evolution Scheme

Jorge Sebastian Hernández-Domínguez[1], Gregorio Toscano-Pulido[1], and Carlos A. Coello Coello[2]

[1] Information Technology Laboratory, CINVESTAV - Tamaulipas, Parque Científico y Tecnológico TECNOTAM. Km. 5.5, Carretera Cd. Victoria-Soto La Marina. Cd. Victoria, Tamaulipas, 87130, México
jhernandez@tamps.cinvestav.mx, gtoscano@cinvestav.mx
[2] CINVESTAV-IPN (Evolutionary Computation Group), Depto. de Computación, Av. IPN No. 2508, San Pedro Zacatenco, México, D.F. 07360, México
ccoello@cs.cinvestav.mx

Abstract. Particle swarm optimization (PSO) and differential evolution (DE) are meta-heuristics which have been found to be very successful in a wide variety of optimization tasks. The high convergence rate of PSO and the exploratory capabilities of DE make them highly viable candidates to be used for solving multi-objective optimization problems (MOPs). In previous studies that we have undertaken [2], we have observed that PSO has the ability to launch particles in the direction of a leader (i.e., a non-dominated solution) with a high selection pressure. However, this high selection pressure tends to move the swarm rapidly towards local optima. DE, on the other hand, seems to move solutions at smaller steps, yielding solutions close to their parents while exploring the search space at the same time. In this paper, we present a multi-objective particle swarm optimizer enhanced with a differential evolution scheme which aims to maintain diversity in the swarm while moving at a relatively fast rate. The goal is to avoid premature convergence without sacrificing much the convergence rate of the algorithm. In order to design our hybrid approach, we performed a series of experiments using the ZDT test suite. In the final part of the paper, our proposed approach is compared (using 2000, 3500, and 5000 objective function evaluations) with respect to four state-of-the-art multi-objective evolutionary algorithms, obtaining very competitive results.

1 Introduction

Particle swarm optimization (PSO) [4] is a meta-heuristic that mimics the behavior of bird flocks by "searching" based on social and personal knowledge acquired by a set of particles. Due to its effectiveness in single-objective optimization, PSO has been extended to multi-objective optimization problems (MOPs). In addition, differential evolution (DE) [6] is another meta-heuristic which has been particularly successful as a single-objective optimizer. DE works by mutating solutions based on the population's variance and this strategy has been found to be a very powerful optimizer in continuous search spaces. Similar

J.-K. Hao et al. (Eds.): EA 2011, LNCS 7401, pp. 169–180, 2012.

to PSO, several DE algorithms have been migrated to multi-objective optimization. Even though both meta-heuristics (PSO and DE) are simple to conceptualize and have shown competitive results on a variety of MOPs, relatively few research has been performed in regards to comparing and contrasting these two meta-heuristics in a multi-objective context. We believe that obtaining more detailed knowledge about the search capabilities of multi-objective evolutionary algorithms (MOEAs) such as these, is of utmost importance to design more powerful algorithms. For example, the few studies currently available indicate that several multi-objective particle swarm optimizers (MOPSOs) have difficulties to deal with multifrontal problems [3], while multi-objective differential evolution (MODE) approaches have shown better results on this type of problems [8,2]. The different behavior of MOPSO and MODE on this type of problems may serve as indicative that a better understanding of these two meta-heuristics will be fruitful.

In this article, we propose a hybrid MOEA that attempts to combine the advantages of MOPSO and MODE. Some of the features adopted for our hybrid approach were obtained from a series of experiments (some of these experiments are included in this document while others were obtained from our previous research [2]) performed on the Zitzler-Deb-Thiele (ZDT) test suite. As a result, aiming to adopt the mechanisms that promote desirable effects (in MODE and MOPSO), we have devised a MOEA which shows competitive results (using the IGD [9] performance measure) when compared to four state-of-the-art MOEAs.

The remainder of this paper is organized as follows. Section 2 introduces some basic concepts related to multi-objective optimization as well as the PSO and DE algorithms. In Section 3, we show a series of experiments that allowed us to better understand the way in which a MODE and a MOPSO algorithms perform the search. Then, in Section 4, we introduce the designed algorithm whose performance is assessed in Section 5. Finally, our conclusions and some possible paths for future research are provided in Section 6.

2   Basic Concepts

In this study, we assume that all the objectives are to be minimized and are equally important. We are interested in solving the general *multi-objective optimization problem* with the following form:

$$\text{Minimize } \mathbf{f}(\mathbf{X}_i) = (f_1(\mathbf{X}_i), f_2(\mathbf{X}_i), \ldots, f_M(\mathbf{X}_i))^T$$
$$\text{subject to } \mathbf{X}_i \in \mathcal{F} \tag{1}$$

where $\mathbf{X}_i$ is a *decision vector* (containing our *decision variables*), $\mathbf{f}(\mathbf{X}_i)$ is the M-dimensional *objective vector*, $f_m(\mathbf{X}_i)$ is the m-th objective function, and $\mathcal{F}$ is the feasible region delimited by the problem's constraints.

2.1   Particle Swarm Optimization

The flight of particles in PSO is typically directed by the following three components: *i) velocity* - this component is conformed by a velocity vector which aids in moving the particle to its next position. Moreover, an inertia weight w is used to control the amount of previous velocity to be applied. ii), **cognitive** - this component represents the "memory" of a particle. This is done with a *personal best* vector (we will refer to this vector as pbest) which remembers the best position found so far by a particle. iii) **social** - this component represents the position of a particle known as the *leader*. The leader is the particle with the best performance on the neighborhood of the current particle. These three components of PSO can be seen on its flying formula. When a particle is about to update its position, a new velocity vector is computed using:

$$\mathbf{v}_i(t+1) = w\mathbf{v}_i(t) + c_1 r_1(t)(\mathbf{y}_i(t) - \mathbf{x}_i(t)) + c_2 r_2(t)(\hat{\mathbf{y}}_i(t) - \mathbf{x}_i(t)) \quad (2)$$

Thereafter, the position of the particle is calculated using the new velocity vector:

$$\mathbf{x}_i(t+1) = \mathbf{x}_i(t) + \mathbf{v}_i(t+1) \quad (3)$$

In Equations (2) and (3), $\mathbf{x}_i(t)$ denotes the position of particle i at time t, $\mathbf{v}_i(t)$ denotes the velocity of particle i at time t, w is the inertia weight, c_1 and c_2 are the cognitive and social factors, respectively, and $r_1, r_2 \sim U(0,1)$. Additionally, $\mathbf{y}_i$ is the best position found so far by particle i (pbest), and, $\hat{\mathbf{y}}_i$ is the neighborhood best position for particle i (leader).

2.2 Differential Evolution

Differential evolution was proposed under the idea that a convenient source for perturbation is the population itself. Therefore, in DE, the step size is obtained from the current population. In this manner, for each parent vector $\mathbf{x}_i$, a difference vector $\mathbf{x}_{i_1} - \mathbf{x}_{i_2}$ is used to perturb another vector $\mathbf{x}_{i_3}$. This can be seen in the following mutation equation,

$$\mathbf{z}_i(t) = \mathbf{x}_{i_3} + F * (\mathbf{x}_{i_1} - \mathbf{x}_{i_2}) \quad (4)$$

where $\mathbf{x}_{i_1}$, $\mathbf{x}_{i_2}$, $\mathbf{x}_{i_3}$ are pairwise different random vectors, and F is a scaling factor.

Recombination in DE is achieved by combining elements from a parent vector $\mathbf{x}_i(t)$ with elements from $\mathbf{z}_i(t)$,

$$\mu_{i,j}(t) = \begin{cases} z_{i,j}(t) \text{ if } U(0,1) < Pr \text{ or } j = r \\ x_{i,j}(t) \text{ otherwise.} \end{cases} \quad (5)$$

where r is a random integer from $[1, 2, \ldots, Dim]$, Pr is the recombination probability and j is used as index for the Dim dimensions of a solution.

3 Analysis of MOPSO and MODE

3.1 Velocity on MOPSO

Most of the current MOPSOs have premature convergence in the presence of multifrontal problems [3]. Some researchers have analyzed the behavior of the velocity of MOPSOs in multifrontal problems and have reported that the reason for their premature convergence is that velocity values grow too much [5,3]. An experiment was developed to study the velocity of MOPSO in more detail. For this experiment, we selected two test problems: ZDT1 (which can be solved relatively easily by a MOPSO), and ZDT4 (which is a multifrontal problem that is very hard to solve for several MOPSOs). For our study, we will adopt OMOPSO[1] which obtained the best results in [3] but still was unable to solve ZDT4. In order to observe the behavior of velocity, we have used boxplots to show the values reached in 30 executions using 50 particles and 100 generations (see Figures 1(a) and 1(b)). In the figures, the y axis shows the velocity values, while the x axis shows the iterations. Please note that the actual boxes is where the majority of observations are located while the dark region (composed of many + symbols) shows the outliers. Figure 1(a) shows that the velocity values for ZDT1 are (approximately) in the range $[-2,2]$. For the case of ZDT4, Figure 1(b) shows that the velocity values are in the range $[-20,20]$. As previously noted by other researchers, the velocity values for ZDT4 are much bigger than for ZDT1. Nonetheless, it should be pointed out that the velocity values shown are proportional to the range of the decision variables of each problem [2]. Moreover, it is important to note that the velocity values in Figure 1(b) fall close to 0 around iteration 60. Since velocity depends on the difference of the leader and the pbest with the current particle's position, we infer that, at some point during the run, particles get really close to the leader and the pbest producing very small velocities. This premise led us to the following experiment.

3.2   Distance of Movement

Our hypothesis is that the observed velocity values are really a consequence, rather than a cause for the premature convergence of OMOPSO in ZDT4. To validate our hypothesis, we will now try to see why is that MODE can achieve much better results on ZDT4. In this experiment we have used the same structure of OMOPSO to design a MODE algorithm. The variant used is $DE/Best/1/Bin$ which showed the best results in previous research. In Figures 2(a) and 2(b), we plotted the Euclidean distance traveled by particles of OMOPSO and solutions of MODE in ZDT4. For OMOPSO, this is the distance between the previous and the new position while in MODE this is the distance between the parent and the candidate position.

Figures 2(a) and 2(b) show different behaviors. OMOPSO reaches bigger distance values than MODE at the beginning of the execution but as the iterations elapse distances in OMOPSO fall rapidly to zero (meaning all particles are landing very close to its previous position). Based on our previous research [2] and from the information seen at the presented plots, we argue that this is due to

---

[1] We have removed the turbulence operator included in this algorithm, since we aim to observe its raw behavior.

[2] ZDT1 has its variables in the range [0,1] while ZDT4 has the first variable in the range [0,1] while the rest are in the range [−5,5].

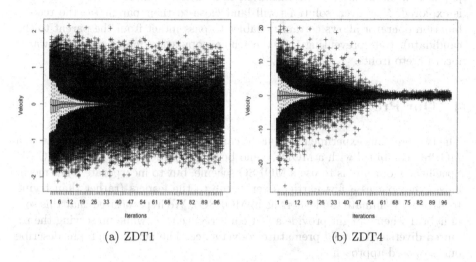

Fig. 1. Velocity of OMOPSO for ZDT1 (left) and ZDT4 (right)

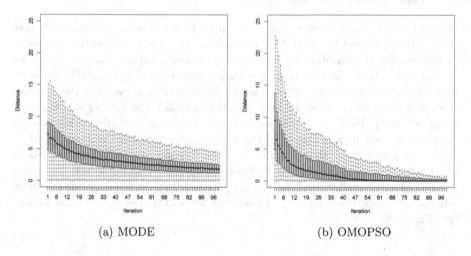

Fig. 2. Euclidean distance traveled from parent to offspring for MODE (left) and OMOPSO (right) on ZDT4

the following: OMOPSO uses only leaders and pbests to move particles while MODE uses information from the entire population. Indeed, in OMOPSO all particles are heavily attracted towards their leaders. If these leaders are diverse enough, then OMOPSO seems to move fast towards improvement. However, if leaders at some point get stuck in a local front, then all solutions will rapidly move towards that front making it harder for OMOPSO to escape (as diversity is very limited). This is not the case for MODE, in which every solution of the population can be used for perturbation and, therefore, more of the search space is explored. Moreover, solutions will land close to their parent (as the recombination operator allows certain variables to pass intact from the parent to the candidate), thus preventing that the whole population moves quickly towards local Pareto fronts.

4   Our Proposal

The two previous experiments led us to conjecture that, if properly integrated, a MOPSO combined with a MODE could be a very powerful yet efficient MOEA. Specifically, our aim is to use a MOPSO scheme, but to incorporate a DE mechanism that places a few particles very close to the leaders (rather than trying to strongly dominate the leaders as in MOPSO). We hypothesized that this sort of hybrid scheme might provide a fast convergence rate, while preserving the required diversity to avoid premature convergence. The next paragraphs describe our proposed approach.

4.1   Leader Selection Scheme

We have developed a scheme which attempts to select a leader that is diverse enough (with respect to the rest of the leaders). Our proposed mechanism works as follows. First, we obtain the centroid of all the available leaders. Then, we use roulette wheel selection to pick a leader such that the probability for each leader to be selected is proportional to the distance of that leader to the centroid (the bigger the distance of the leader to the centroid, the greater its chance of being selected). We believe that this scheme should be able to provide enough diversity in cases in which most of the leaders move towards the same region.

Here, one can argue that a leader selection scheme based on crowding would give similar results to ours. There is, however, an important difference. Crowding will favor both boundaries of the search space at all times, whereas our proposed scheme will favor the boundary opposite to the location of most of the leaders (see Figure 3).

4.2   The Use of the Velocity

It has been a common practice in PSO to decrease the previous velocity using a factor w. Moreover, this factor has traditionally been set using one of three schemes: i) w adopts a constant value, ii) w depends on the current iteration, and iii) w is randomly selected from a range. Here, we have adopted a different scheme in which w depends on the previous success of the particle. We believe that if a particle succeeded at previous iterations, then it should be beneficial to use a bigger portion of its previous velocity. On the contrary, if the particle has not been successful, then the previous velocity portion should be decreased giving a higher chance to the social and cognitive factors to take action.

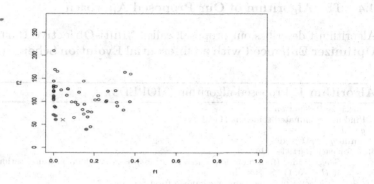

Fig. 3. Leader selection scheme for our proposal. Leaders are represented by filled dots, solutions are black circles and the "X" symbol is the centroid of the leaders

In short, Equation (6) works in the following way[3]. If a particle is the selected leader (meaning the particle has found a very good position at its previous iteration), then we use all of its previous velocity ($w = 1$). If a particle's current position is its pbest (the particle found its best position of the whole execution in the previous iteration), then we use a high range of its previous velocity ($w = U \sim (0.5, 1.0)$). In none of the two previous cases occur, then we use a lower percentage of the velocity ($w = U \sim (0.1, 0.5)$).

$$w = \begin{cases} 1 & \text{if } \mathbf{x}_i(t) = \hat{\mathbf{y}}_i(t) \text{ and } flip(0.5) = true \\ U \sim (0.5, 1.0) & \text{if } \mathbf{x}_i(t) = \mathbf{y}_i(t) \text{ and } flip(0.5) = true \\ U \sim (0.1, 0.5) & \text{otherwise.} \end{cases} \quad (6)$$

4.3   Moving towards an Specific Leader

We adopt a DE mechanism to place some particles close to the selected leader rather than finding a position that strongly dominates the leader but might be moving "deeper" into some local optima. Thus, using a low probability we adopt a mechanism which will make several variables (with a high probability) to be equal to the selected leader. The aim is that the obtained particle will be close to the selected leader and that region of the search space is not lost. Moreover, a few variables will be obtained from a mutation based on three different random pbests. Equation (7) describes this mechanism.

$$\mathbf{x}_{i,j}(t) = \begin{cases} \mathbf{y}_{i_3,j} + F * (\mathbf{y}_{i_1,j} - \mathbf{y}_{i_2,j}) & \text{if } U(0,1) < Pr \text{ or } j = r \\ \hat{\mathbf{y}}_{p,j} & \text{otherwise.} \end{cases} \quad (7)$$

---

[3] $flip(0.5)$ refers to a function which returns true with 50% probability.

4.4 The Algorithm of Our Proposed Approach

Algorithm 1 describes our proposal, called **Multi-Objective Particle Swarm Optimizer Enhanced with a Differential Evolution Scheme (*MOPEDS*)**.

Algorithm 1. Proposed algorithm (MOPEDS)

Initialize Population
Find non-dominated solutions (Leaders)
$g = 0$
while $g < gMax$ **do**
   for each Particle i **do**
      Select leader (from non-dominated solutions) using mechanism from Section 4.1
      if $U \sim (0,1) > Pm$ **then**
         Select a w value using mechanism from Section 4.2
         Update velocity
         Update position
      else
         Select three random (different) pbest
         Move particle using DE scheme from Section 4.3
      end if
      Evaluate particle i
      if Particle i position dominates its pbest **then**
         update pbest to current position
      end if
   end for
   Update non-dominated solutions (Leaders)
end while

---

5 Experimental Study

Next, we compare our proposal with four state-of-the-art MOEAs: OMOPSO[4] [3], SMPSO [5], NSGA-II[5] [1], and DEMO [6] [8]. The following parameters were adopted. For NSGA-II: 0.9 for the crossover rate, $1/Dim$ for the mutation rate, 15 for the distribution index for crossover, and 20 for the distribution index for mutation. DEMO uses $Pr = 0.3$, and $F = 0.5$. OMOPSO uses $C1, C2 = U \sim (1.5, 2.0)$, and $w = U \sim (0.1, 0.5)$. Finally, SMPSO uses $C1, C2 = U \sim (1.5, 2.5)$, and $w = U \sim (0.1, 0.5)$. MOPEDS adopted $Pm = 0.2$, $F = G(0.5, 0.5)$, $C1, C2 = U \sim (1.2, 2.0)$ and $Pr = 0.2$, since these parameters provided the best behavior in our preliminary tests. The w parameter is used as described in Section 4.2. We have compared all algorithms with respect to the IGD performance measure using 2000, 3500, and 5000 objective function evaluations in order to obtain more detailed information about their behavior. There is one plot for each problem in which each algorithm is presented at the three different numbers of function evaluations indicated above (see Figure 4).

    ZDT1 - In this problem it can be observed that MOPEDS is ahead of all the other algorithms at 2000 function evaluations. This shows the fast convergence

---

[4] The implementation of OMOPSO used in our experiments differs from its original proposal [7] in that ϵ-dominance is not adopted.

[5] We took the code available at: `http://www.iitk.ac.in/kangal/codes.shtml`

[6] Please do not confuse DEMO with MODE. DEMO is an specific implementation of MODE which refers to multi-objective differential evolution.

rate of our proposal on this problem. Moreover, at 3500 evaluations MOPEDS is still ahead and DEMO is just a little behind. Finally, when reaching 5000 evaluations MOPEDS and DEMO show IGD values very close to 0 (the ideal value).

ZDT2 - Again, MOPEDS has the best performance at 2000 function evaluations. In fact, MOPEDS is capable of obtaining considerable advantage over the rest of the algorithms in this number of evaluations. At 3500 evaluations, IGD values for MOPEDS are already very close to 0 while OMOPSO and DEMO are a little behind. Finally, at 5000 function evaluations, OMOPSO and DEMO have reached values as good as MOPEDS.

ZDT3 - MOPEDS shows the best IGD values at 2000 and 3500 evaluations. Nonetheless, at 3500 evaluations, DEMO is very competitive also. Furthermore, at 5000 evaluations MOPEDS and DEMO have a very similar performance.

ZDT4 - In this problem, it is clear the SMPSO shows much better results than any other algorithm, while OMOPSO shows the poorest. Acknowledging this, we will omit OMOPSO and SMPSO from the discussion of this problem. At 2000 evaluations, our proposal is ahead of DEMO and a little bit behind NSGA-II. At 3500 evaluations, MOPEDS and NSGA-II show similar results. Nonetheless, the box is bigger for MOPEDS indicating a bigger dispersion than NSGA-II. Finally, at 5000 evaluations, our proposal is again competitive with NSGA-II.

ZDT6 - It is clear that all algorithms (except one) get close to the true Pareto front very early in the search (after only 2000 function evaluations). In this problem, NSGA-II requires a larger number of function evaluations to reach the true Pareto front.

Since SMPSO achieved such excellent results in ZDT4, we decided to analyze this algorithm in more detail. SMPSO is actually a modification of OMOPSO in which a velocity constriction mechanism is used[7]. Basically, this constriction factor limits the values that the velocity can take before moving a particle. The velocity is limited using Equations (8) and (9):

$$v_{i,j} = \begin{cases} delta & \text{if } v_{i,j} > delta_j \\ -delta_j & \text{if } v_{i,j} \leq -delta_j \\ v_{i_j} & \text{otherwise.} \end{cases} \tag{8}$$

$$delta_j = \frac{(upper\_limit_j - lower\_limit_j)}{2} \tag{9}$$

This constriction factor is the main difference with respect to the original OMOPSO. After analyzing the behavior of this approach in ZDT4, we reached the following conclusions. For ZDT4 (from the second to the tenth variables) $\delta = 5$ and $-\delta = -5$. Moreover, SMPSO (as many other MOEAs) truncates variables to their upper and lower limits when these go beyond their predefined bounds. For ZDT4, these bounds are -5 and 5. Therefore, if a variable at iteration t lands above its upper limit, this variable will be truncated to 5. Then, at

---

[7] Please refer to [5] for further details on this algorithm.

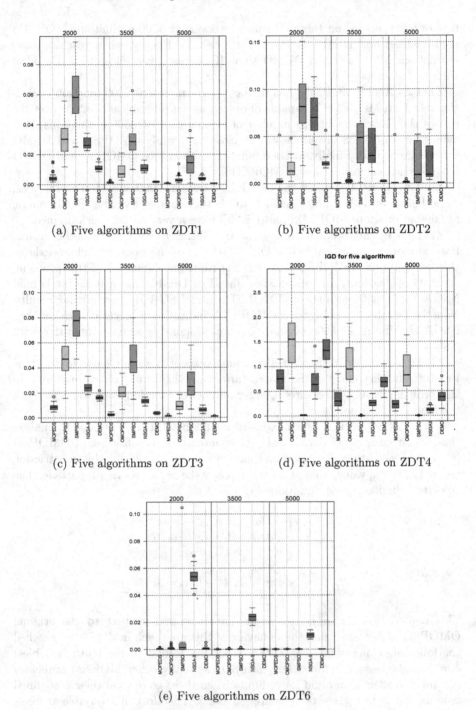

(a) Five algorithms on ZDT1

(b) Five algorithms on ZDT2

(c) Five algorithms on ZDT3

(d) Five algorithms on ZDT4

(e) Five algorithms on ZDT6

Fig. 4. Comparison of 5 algorithms using the IGD performance measure at 2000, 3500, and 5000 function evaluations in the ZDT test suite

iteration $t + 1$, if it happens that the velocity goes below -5 (which is $-\delta$), this velocity will be truncated to $-\delta = -5$. Therefore, when we add the velocity to the current particle's position $5 + (-5)$, we end up with 0 which is precisely in the region where the Pareto optimal set for this test problem is located. Even when this is a very clever mechanism and works perfectly in ZDT4, we decided to observe the robustness of SMPSO when the ranges of ZDT4 are changed. Thus, we moved the lower limit (again from the second to the tenth variable) to -2. The upper limit was not changed. Moreover, we also tested our proposal using these modifications with ZDT4. Both algorithms were run using 25000 function evaluations (100 particles and 250 iterations). Results are shown in Figure 5.

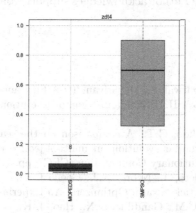

Fig. 5. MOPEDS and SMPSO at a modified version of ZDT4, using $25,000$ evaluations

From Figure 5, we can observe that the performance of SMPSO is clearly deteriorated when changing the ranges of the variables for the ZDT4 problem. We can see that SMPSO works very well some times while reporting poor results at other executions. Regarding our proposal, we observe that the modification did not have a significant impact on its performance. In fact, some improvements were achieved with regards to its execution at 2000, 3500, and 5000 evaluations. It is important to note, however, that our proposal was never able to reach the true Pareto front but could only closely approximate it.

6   Conclusions and Future Work

In conclusion, our proposal shows competitive results when compared with other state of the art MOEAs using a small number of function evaluations. Moreover, it is important to note that our proposal is able to reach competitive results in the multifrontal problem ZDT4 which has found to be quite difficult for most current MOPSOs. Therefore, we believe this indicates that MOPEDS is benefitting from the high convergence rate of MOPSO while maintaining diversity using the DE scheme. It is important to note, however, that the number of evaluations adopted (except for 25000) are relatively small and, therefore, the plots presented here do not give further information about the performance of MOPEDS if we extend its execution. This seems important since our algorithm was not able to reach the true Pareto front at 25000 evaluations. This is certainly an issue that deserves some further work. Finally, we also believe that further investigation on mechanisms to fine tune the parameters of our proposed approach are a promising research path.

Acknowledgements. The first author acknowledges support from CONACyT through a scholarship to pursue graduate studies at CINVESTAV-Tamaulipas. The second author gratefully acknowledges support from CONACyT through project 105060. The third author acknowledges support from CONACyT project no. 103570

References

1. Deb, K., Pratap, A., Agarwal, S., Meyarivan, T.: A Fast and Elitist Multiobjective Genetic Algorithm: NSGA–II. IEEE Transactions on Evolutionary Computation 6(2), 182–197 (2002)
2. Dominguez, J.S.H., Pulido, G.T.: A comparison on the search of particle swarm optimization and differential evolution on multi-objective optimization. In: 2011 IEEE Congress on Evolutionary Computation (CEC), pp. 1978–1985 (June 2011)
3. Durillo, J.J., García-Nieto, J., Nebro, A.J., Coello Coello, C.A., Luna, F., Alba, E.: Multi-Objective Particle Swarm Optimizers: An Experimental Comparison. In: Ehrgott, M., Fonseca, C.M., Gandibleux, X., Hao, J.-K., Sevaux, M. (eds.) EMO 2009. LNCS, vol. 5467, pp. 495–509. Springer, Heidelberg (2009)
4. Kennedy, J., Eberhart, R.: Particle Swarm Optimization. In: IEEE International Conference on Neural Networks, vol. 4, pp. 1942–1948 (1995)
5. Nebro, A.J., Durillo, J.J., Garcia-Nieto, J., Coello Coello, C.A., Luna, F., Alba, E.: SMPSO: A New PSO-based Metaheuristic for Multi-objective Optimization. In: 2009 IEEE Symposium on Computational Intelligence in Multi-Criteria Decision-Making (MCDM 2009), March 30-April 2, pp. 66–73. IEEE Press, Nashville (2009) ISBN 978-1-4244-2764-2
6. Price, K., Storn, R.: Differential Evolution - a simple evolution strategy for fast optimization (April 1997)
7. Reyes Sierra, M., Coello Coello, C.A.: Improving PSO-Based Multi-objective Optimization Using Crowding, Mutation and ε-Dominance. In: Coello Coello, C.A., Hernández Aguirre, A., Zitzler, E. (eds.) EMO 2005. LNCS, vol. 3410, pp. 505–519. Springer, Heidelberg (2005)
8. Robič, T., Filipič, B.: DEMO: Differential Evolution for Multiobjective Optimization. In: Coello Coello, C.A., Hernández Aguirre, A., Zitzler, E. (eds.) EMO 2005. LNCS, vol. 3410, pp. 520–533. Springer, Heidelberg (2005)
9. Veldhuizen, D.A.V., Lamont, G.B.: On Measuring Multiobjective Evolutionary Algorithm Performance. In: 2000 Congress on Evolutionary Computation, vol. 1, pp. 204–211. IEEE Service Center, Piscataway (2000)

Evolution of Multisensory Integration in Large Neural Fields

Benjamin Inden[1], Yaochu Jin[2], Robert Haschke[3], and Helge Ritter[3]

[1] Research Institute for Cognition and Robotics, Bielefeld University
binden@cor-lab.uni-bielefeld.de
[2] Department of Computing, University of Surrey
[3] Neuroinformatics Group, Bielefeld University

Abstract. We show that by evolving neural fields it is possible to study the evolution of neural networks that perform multisensory integration of high dimensional input data. In particular, four simple tasks for the integration of visual and tactile input are introduced. Neural networks evolve that can use these senses in a cost-optimal way, enhance the accuracy of classifying noisy input images, or enhance spatial accuracy of perception. An evolved neural network is shown to display a kind of McGurk effect.

1 Introduction

Animals take advantage of different sensory facilities to react appropriately to their environment. Among these facilities are visual, auditory, tactile, olfactory, gustatory, and electric senses. Researchers are typically interested in questions like: How are percepts from different modalities perceived as a single entity? What happens if there is conflicting input from different modalities? How can top-down influences (attention) bias the process of integration? [1,2]

Here we are particularly interested in evolutionary aspects of multisensory integration. Advantages of multisensory integration may include increased spatial accuracy of perception, increased classification accuracy, and minimizing costs associated with different forms of perception. The purpose of this study is to introduce a number of sufficiently understandable tasks that require multisensory integration to be solved, to verify that artificial neural networks can evolve to solve these tasks, and to get a rough look at how these networks do it.

Studying the integration of input from different modalities has already been possible for some time in the context of evolutionary robotics. One important difference between earlier studies and what we present here is that the techniques for evolving neural networks have advanced significantly in the meantime [3]. Previously, only a few contact or distance sensors were typically used. If there was camera input, it was usually pre-processed to reduce its dimension [4]. It is now possible to evolve very large neural networks that take input from not just a few sensors, but from whole sensor arrays, e.g. cameras. The NEAT-fields method used here for neuroevolution is particularly suited to these tasks

J.-K. Hao et al. (Eds.): EA 2011, LNCS 7401, pp. 181–192, 2012.

because it evolves not just the connectivity between single neurons, but also the connectivity between complete neural fields. How this method solves a range of different problems and how its implementation details and default parameter settings were determined is discussed in detail elsewhere [5,6,7]. We expect that some other recent neuroevolution methods could also solve the tasks presented here. A comparison between NEATfields and these other methods has been done previously [7] and confirms that using NEATfields here is a reasonable choice. Of course, multisensory integration has also previously been studied using large hand designed networks (e.g. [8]). However, an advantage of evolving neural networks is that often solutions are found that are free from human bias, or beyond human imagination.

2   Methods

2.1   The NEATfields Method

Neural Networks. NEATfields is based on the well known NEAT method [9] that simultaneously evolves the topology and the connection weights of neural networks. NEATfields networks have three levels of architecture (see fig. 1 (a)). At the highest level, a network consists of a number of fields that are connected just like individual neurons are in a NEAT network. At the intermediate level, fields are collections of identical (or similar) subnetworks with a two dimensional topology. At the lowest level, these subnetworks, termed field elements, are recurrent NEAT networks composed of individual neurons. A complete NEATfields network consists of at least one internal field as specified by the genome, and

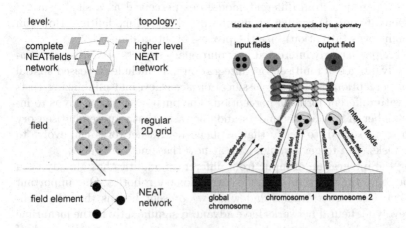

Fig. 1. (a) The three levels of architecture in NEATfields. (b) Construction of a NEATfields network from its genome (bottom). The balls in the central network represent field elements. Their contents are shown in the circles above and below. In these circles, black dots represent the individual (input, output or hidden) neurons.

fields for network input and output as specified by the given task. There can be several input and output fields with different dimensions. For example, a bias input can be provided as an input field of size 1×1. Within the NEATfields network, connections can be local (within a field element), lateral (between field elements of the same field), or global (between two fields). It should be noted that connections between field elements or fields are in fact connections between individual neurons in these field elements or fields. How they are established will be described below. Evolution is started with a single internal field of size 1×1 that is connected to all input and output fields. So there is initially only one field element, and it contains one neuron for each different output.

The activation of the neurons in NEATfields is a weighted sum of the outputs of the neurons $j \in J$ to which they are connected, and a sigmoid function is applied on the activation: $o_i(t) = \tanh(\sum_{j \in J} w_{ij} o_j(t-1))$. Connection weights are constrained to the range $[-3, 3]$. There is no explicit threshold value for the neurons. Instead, a constant bias input is available in all networks.

Encoding. The parameters for an individual field are encoded in the genome on a corresponding chromosome (see fig. 1 (b)). The first gene in a chromosome specifies the field size in x and y dimensions. After that node and connection genes for one field element are specified. All genes contain a unique reference number that is assigned once the gene is generated through a mutation. In addition, connection genes contain a connection weight, a flag indicating whether the connection is active, and the reference numbers of the source and target neurons.

There is a special chromosome that contains genes coding for global connections. Global connections contain reference numbers of a source and a target. These can be reference numbers either of neurons or of inputs and outputs as

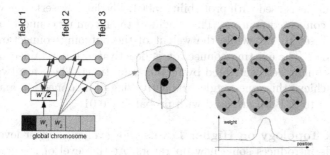

Fig. 2. (a) An example of how the global connections between neurons in fields of different sizes (shown here as one dimensional) are created using a deterministic and topology preserving method. The genetically specified weights are automatically scaled if necessary. As shown in detail on the right side, connections go in fact to individual neurons within the field elements as specified by the connection gene. (b) Dehomogenization of neural fields by the focal areas technique. The thickness of connections here symbolizes their weights.

specified by the task description. They must be in different fields — global connections between two neurons in the same field are never created by the NEAT-fields method. Due to the higher level architecture of NEATfields, a neuron with a given reference number will be present n times in a field with n field elements. A single global connection gene implicitly specifies connections for all these neurons. If a global connection is between neurons in fields with the same sizes, every field element in the target field will get a connection from the field element in the source field that has the same relative position (in x and y dimension) in its field. Their connection weights are all the same because they are all derived from one gene. If field sizes are different in a dimension, then the fields will still be connected using a deterministic and topology preserving method (see fig. 2 (a)): if the source field is smaller than the target field, each source field neuron projects to several adjacent target field neurons, whereas if the source field is larger than the target field, the target field neurons get input from a number of adjacent source field neurons, while the genetically specified connection weight is divided by that number. That way, field sizes can mutate without changes in the expected input signal strength.

Mutation Operators for the Field Elements. The NEATfields method uses mutation operators that are very similar to those of the original NEAT implementation for evolving the contents of the field elements. The most common operation is to choose a fraction of connection weights and either perturb them using a normal distribution with standard deviation 0.18, or (with a probability of 0.15) set them to a new value. The application probability of this weight changing operator is set to 1.0 minus the probabilities of all structural mutation operators, which amounts to 0.8815 in the experiments reported here. In general, structural mutations are applied rarely because they will cause the evolutionary process to operate on larger neural networks and search spaces. A structural mutation operator to connect neurons is used with probability 0.02, while an operator to insert neurons is used with probability 0.001. The latter inserts a new neuron between two connected neurons. The weight of the incoming connection to the new neuron is set to 1.0, while the weight of the outgoing connection keeps the original value. The existing connection is deactivated but retained in the genome where it might be reactivated by further mutations. There are two operators that can achieve this: one toggles the active flag of a connection and the other sets the flag to 1. Both are used with probability 0.01.

Evolving Network Topology on Higher Levels. For evolving higher level topology, NEATfields introduces some new operators. At the level of a single field, one operator doubles the field size along one dimension (with a probability of 0.001) and another changes the size (for both dimensions independently) to a random value between its current size and the size of the largest field it is connected to (with a probability of 0.005). NEATfields networks can also have lateral connections between field elements of the same field. Lateral connections are like local connections: they are between two neurons in the NEAT network that describes the field elements. However, the connection from the source

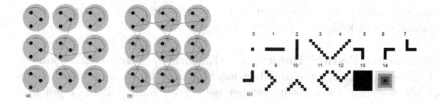

Fig. 3. (a) Lateral connections are established between a neuron in a field element and another neuron in each of the up to four neighbor field elements. There are less neighbors if it is at the border of the field. Lateral connections are only shown for the central element as dotted lines here for clarity. (b) All lateral connections constructed from a single gene. (c) Hand-designed patterns for local feature detectors. Black denotes a connection weight of 1.0, white a connection weight of -1.0, and the gray scales denote connection weights between 0.0 and 1.0.

neuron does not go to a neuron in the same field element, but to the corresponding neurons in the up to four neighbor field elements instead (see fig. 3). The gene coding for a lateral connection specifies source and target neuron reference numbers just as genes coding for local connections do; it is also located in the same chromosome. However, it has a lateral flag set to 1, and is created by a lateral connect operator (with a probability of 0.02).

By default, corresponding connections in different field elements all have the same strength so they can be represented by one gene. The same is true for the global connections between field elements of two fields. For some tasks, it is useful to have field elements that are slightly different from each other. One way to realize this is to make a connection weight larger in a neighborhood of some center coordinates on the field. The connection weight is scaled according to $\exp(-\epsilon(\frac{distance}{field\,size})^2)$ (in our implementation, this is done separately for the x and y dimensions), where $\epsilon = 5.0$ (see figure 2 (b)). The center coordinates are specified in the genome. There is a mutation operator that (at a probability of 0.03) sets these properties for a single connection gene.

At the level of the complete NEATfields network, there is an operator that inserts global connections (with a probability of 0.01) and an operator that inserts a new field into an existing global connection (with a probability of 0.0005). The size of the new field is set randomly to some value between 1 and the larger of the sizes of the two fields between which it is inserted. This is done independently for both dimensions. An existing field can also be duplicated, where all elements of the new field receive new reference numbers. The new field can either be inserted parallel to the old one (in this case, the outgoing connection weights will be halved to prevent disruption of any existing function) or in series with the old one (in this case, every neuron in every field element in the new field gets input from the corresponding neuron in the corresponding field element in the old field, while the output from the new field goes to where the output from the old field went previously). The serial duplication operator is also applied on input fields, in which case an internal field is created that contains one neuron for every input in the input field. Both mutations occur with probability 0.0005.

Building Local Feature Detectors. NEATfields also has another kind of global connections that serve to detect local patterns by connecting to a number of adjacent field elements in another field. For example, if the input field contains camera data, they can detect lines or edges in particular orientations. These global connections have an additional number on the connection gene that refers to one of at most 16 prespecified connection weight patterns. The patterns have been designed by hand with the aim of providing useful building blocks for local feature detection (see fig. 3 (c)). A value from the pattern is multiplied by the original connection weight specified in the gene to get the weight of the connection from the field element at the respective position. The patterns have a size of 7×7. The central element of the pattern corresponds to the weight modification of the connection between exactly corresponding field elements. The feature detectors do not have to use an entire pattern. Instead, the maximal offsets between the field element positions in the source and target fields can be between 0 and 3 separately for the x and y dimensions, and are also specified on the genome. There is a mutation operator (used with a probability of 0.02) that mutates the pattern reference and maximal offsets of an already existing connection. Here we use only patterns 1 to 4 from figure 3 (c) with equal probabilities, and the maximal offsets in x and y direction are 0 or 1 with equal probabilities. Comparisons to other parameter settings and a number of alternative approaches can be found in earlier work [7].

2.2   Selection Methods

We use speciation selection based on genetic similarity as described previously [7] for the spatial accuracy task. For the other three tasks, we use a hybrid selection method that combines fitness based tournament selection and a recent approach known as "novelty search". Half of the population is chosen based on fitness, and the other half is chosen based on novelty. Each half has its own elite of size 10 that is copied to the next generation unchanged. This method has also been described elsewhere [6]. The behavioral novelty measures used are described in the sections that describe the respective task to be solved. In all cases, a population of 1000 individuals is used.

3   Tasks

3.1   The "Scan/Lift" Task

This classification task can be optimally solved by integrating information from different sensory channels. The network sees a different image of size 11×11 in each of the four episodes (see fig. 4a). These patterns have to be assigned to two

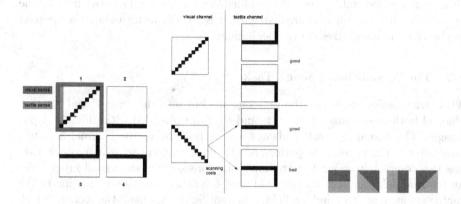

Fig. 4. Input patterns for multisensory integration tasks. (a) The Scan/Lift task. There are four episodes in this task with four different images. The visual sense sees the whole image after a lift action was executed, whereas the tactile sense can scan the image column-wise. (b) The Conditional Scan task. There are four episodes with different combinations of visual and tactile input. Tactile scanning is only required in the lower two episodes to disambiguate the input. (c) The four classes of gray scale images presented to neural networks in the third task.

classes depending on whether the line goes up or down on the right side of the image. The neural network has four outputs: one classification output for each class, and two outputs for active perception. Every 10 time steps, the output with the highest value is considered to be an action. If it is a classification action, the episode terminates, and the reward is determined. The active perception outputs control one of the two sensory modalities each. The first controls the "visual" modality. If it is active, a "lift" command is executed, which means the whole pattern becomes visible in the visual input field. At the beginning of an episode, the pattern is not visible, so all visual inputs are 0.0. The second active perception input controls the "tactile" input, which can look at the same image in a different way. If it is active, a "scan" command is executed, which means that the scanning apparatus (which provides a single column of the image as input) moves one column to the right. At the beginning of an episode, it shows the leftmost column. If it has already been moved to the rightmost column, it will not move further.

The neural network gets 35 fitness points for every correct classification. In addition, it gets $15 - 5a_L - a_S$ fitness points in every episode, where a_L denotes whether the lift action was invoked (0 or 1), and a_S how often the scan action was invoked. So lifting is more expensive than scanning a row, but it is cheaper than scanning the whole image. This means that a network can get the optimal fitness if it uses scanning once for images 1 and 2 (where the line begins in the bottom of the image), but lifting for images 3 and 4 (where the line begins in the middle of the image). In that case, it gets a total fitness of 188 points. The behavioral distance between individuals is calculated as follows: If the corresponding commands of the two individuals at time step t are unequal, $\frac{10}{t+1}$ points are added to the distance. Times where only one of the individuals is still alive are disregarded. So this measure assigns the greatest weights to commands given at the beginning.

3.2   The "Conditional Scan" Task

This task (see fig. 4b) is like the previous one but now the "visual" and "tactile" channel both have a size of 11×11, and they are provided with different input images. The visual channel is always active, but the tactile channel is active only after a tactile scan is performed. Evolution starts with a network that has two completely separate partitions, one consisting of the visual input field, an internal field, and the output field, and the other partition consisting of the tactile channel, another hidden field, and another output field, the state of which is ignored. Evolution can subsequently add more fields or connect them. There are three possible actions for the network: scan, assign to the "good" class, or assign to the "bad" class. "Scan" makes the whole image appear at once on the tactile channel. The neural network gets 10 fitness points for correct classification of a "bad" pattern, 5 points for correct classification of a "good" pattern, and 1 point if the scan action was not used in an episode. So the highest obtainable fitness is 27. Behavioral distance is calculated as for the previous task.

3.3   A Multisensory Classification Task

In this task the networks have a visual input channel of size 11×11. 32 patterns of 4 different classes are displayed in separate episodes (see fig. 4c). However, the patterns are very noisy. The pixels in the area with the brighter color are generated from a normal distribution with mean 0.1 and variance 0.3, while those in the darker area are generated from a normal distribution with mean -0.1 and variance 0.3. The network has four classification outputs and an output to generate a scan command. Its tactile sense is of size 11×1 and looks at a single row of another image of the same class. Initially, the scanner is at position $x = -1$, so no input is perceived. Scanning does not incur any costs. There is a control condition were all rows are empty (i.e., the tactile input is 0.0). The neural network gets 1.0 fitness points for every episode in which it classifies the image correctly. If the classification is wrong, the network can get up to 0.5 fitness points according to $f = 0.5 - 0.25(o_a - o_c)$, where o_a is the highest output activation, and o_c is the activation of the correct output. Behavioral distance is calculated as for the other tasks.

3.4   Increasing Spatial Accuracy by Sensorimotor Integration

In this task 64 randomly generated images each containing a spatial cue are presented to a network in separate episodes. The input value is 1.0 at the position of the cue, and 0.0 elsewhere. The network can perceive the images through a visual and a tactile channel, each of size 11×11. The positions of the cues on

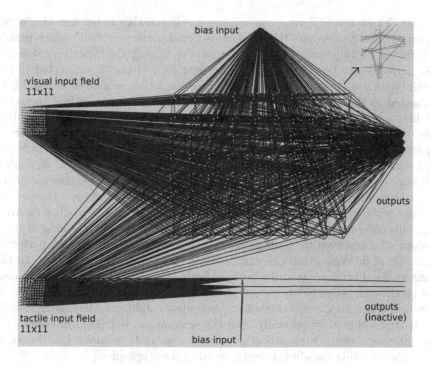

Fig. 5. A neural network solving the "Conditional Scan" task. In the upper left corner, a field element of the large internal field is shown magnified.

the input fields randomly deviate from the true position of the cue by up to ± 2 in both dimensions. For both input fields there is each an internal field and an output field. These two pathways are completely separated in the common ancestor. However, the network can gain higher fitness if it has evolved connections to average over the positions of the cues in the two input fields. The output fields are also of size 11×11. In each episode, the network gets a reward of $2 - (d(c, o_1)^2 + d(c, o_2)^2)/s^2$, where d is Euclidean distance, c is the true position of the cue, and the o_i are the positions on the two output fields with the highest activation. Behavioral distance between two individuals is defined as the sum over all episodes of the squared distances between the positions with highest activation on the corresponding output fields.

4 Results

20 runs of each of the four tasks are performed. The "Scan/Lift" task is solved perfectly in 75% of the runs using 349883 evaluations on average. In 15% of the runs, the classification is done correctly by the winner, but the "lift" action is used in all episodes, which results in a fitness of 180 points only. In 10% of the runs, the fitness is between 180 and the maximum 188 points. The "Conditional Scan" task is solved perfectly in 85% of the runs using 329705 evaluations on

average. The other 15% runs converge on a solution that never uses the "scan" action and always classifies the ambiguous visual pattern as "bad", which results in a fitness of 24 points. One of the networks solving the "Conditional Scan" task perfectly can be seen in fig. 5. The internal field that is connected to the visual input has grown to a size of 10×8 by evolution, while the internal field connected to the tactile input has not grown. Instead, evolution has connected the first internal field to the tactile input as well. It can be seen that evolution uses one recurrent connection within each field element. Further recurrence is achieved through mutual lateral connections between some neurons in different field elements. It should be noted that solutions from other runs look very much different. In particular, some solutions for these tasks use several internal fields.

In the multisensory classification task, 31.55 fitness points are reached on average, while in its unisensory control, 30.74 fitness points are reached on average. This difference is slight but significant ($p < 0.001$, t-test). Looking at the 9 evolved solutions from the multisensory condition that solved the task perfectly, a number of different strategies can be found. One solution does not use tactile scans at all. Therefore, it still works perfectly when it is evaluated under control conditions. Seven solutions use tactile scans in between 8 and 28 of the 32 episodes. When they are evaluated under control conditions, they never classify the patterns that they normally used the scanner on, but perform scans until their lifetime is over. One solution uses the scanner in 17 episodes. Under control conditions, it stills classifies correctly in 10 of these episodes.

In the spatial accuracy task, 125.1 fitness points are reached on average, while in its unisensory control, 123.4 fitness points are reached on average. This is a significant difference ($p < 10^{-7}$, t-test). When having a closer look at the evolved network with the highest fitness, we found that it displayed a kind of McGurk effect. This effect originally refers to the perception of an intermediate phoneme when a subject simultaneously sees a video of a person producing one

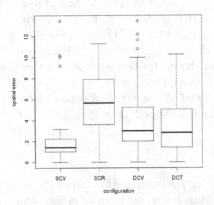

Fig. 6. Errors in the spatial accuracy task for four configurations: SCV - same cue condition / visual cue; SCR - same cue condition / suppressed cue; DCV - different cue condition / visual cue; DCT - different cue condition / tactile cue. All differences, except between DCV and DCT, are significant

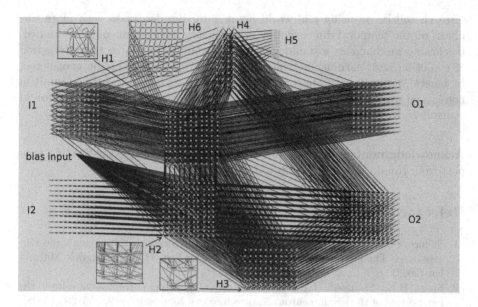

Fig. 7. An evolved network for the spatial accuracy task. "I" denotes an input field, "H" a hidden/Internal field, and "O" an output fields. The fine structure of three internal fields is shown in magnification. Field H4 only produces constant output, while fields H5 and H6 do not produce outputs because the respective connections have been deactivated by evolution. Multisensory integration is achieved by mutual connections between H1 and H2. Furthermore, H3 processes inputs from both sensor fields, but only contributes output for O2.

phoneme and hears a recording of that person producing another phoneme. It demonstrates that speech perception is based on multimodal information [1]. Our network was examined under two conditions. The first was to provide it with a new set of 100 randomly chosen locations and record the distances to the cues and the distances to another set of randomly chosen cues that were not displayed. The second condition was to test it with the same 100 locations again, but this time the tactile cues were from the other random set, so they were different from the visual cues. The distance to both cues was recorded. The network displayed a kind of McGurk effect because the distances to the cues in the second condition both were significantly larger than the distance to the cue in the first condition, and were both significantly smaller than the distance to the non-visible cue in the first condition (see fig. 6) That means that the response was intermediate between both cues. The network is shown in fig. 7.

5   Discussion

We have shown that multisensory integration can evolve for cost-optimal use of sensors or increased classification or spatial accuracy. Importantly, this has been done not just using a few sensors, but using complete images. Further research will be on using real world input data. Besides, tasks with more emphasis on the temporal dimension on multisensory integration could be studied. If learning mechanisms are added to neuroevolution methods, the interaction of evolution and learning in multisensory integration can also be considered. In general, evolved neural networks that can process high-dimensional input in non-trivial ways are not just relevant for the goal of automatic construction of controllers, but also provide a useful tool for neuroscience research [10].

Acknowledgments. Benjamin Inden gratefully acknowledges the financial support from Honda Research Institute Europe.

References

1. Spence, C.: Multisensory integration, attention and perception. In: Signals and Perception — The Fundamentals of Human Sensation, pp. 345–354. Palgrave Macmillan (2002)
2. Stein, B.E., Stanford, T.R.: Multisensory integration: current issues from the perspective of the single neuron. Nature Rewievs Neuroscience 9, 255–266 (2008)
3. Floreano, D., Dürr, P., Mattiussi, C.: Neuroevolution: from architectures to learning. Evolutionary Intelligence 1, 47–62 (2008)
4. Nolfi, S., Floreano, D.: Evolutionary Robotics — The Biology, Intelligence, and Technology of Self-Organizing Machines. MIT Press (2000)
5. Inden, B., Jin, Y., Haschke, R., Ritter, H.: Neatfields: Evolution of neural fields. In: Proceedings of the Conference on Genetic and Evolutionary Computation (2010)
6. Inden, B., Jin, Y., Haschke, R., Ritter, H.: How evolved neural fields can exploit inherent regularity in multilegged robot locomotion tasks. In: Third World Congres on Nature and Biologically Inspired Computation (2011)
7. Inden, B., Jin, Y., Haschke, R., Ritter, H.: Evolving neural fields for problems with large input and output spaces. Neural Networks 28, 24–39 (2012)
8. Westermann, G.: A model of perceptual change by domain integration. In: Proceedings of the 23rd Annual Conference of the Cognitive Science Society (2001)
9. Stanley, K., Miikkulainen, R.: Evolving neural networks through augmenting topologies. Evolutionary Computation 10, 99–127 (2002)
10. Ruppin, E.: Evolutionary autonomous agents: A neuroscience perspective. Nature Reviews Neuroscience 3, 132–141 (2002)

Reducing the Learning Time of Tetris in Evolution Strategies

Amine Boumaza

Univ. Lille Nord de France, F-59000 Lille, France,
ULCO, LISIC, F-62100 Calais, France
boumaza@lisic.univ-littoral.fr

Abstract. Designing artificial players for the game of Tetris is a challenging problem that many authors addressed using different methods. Very performing implementations using evolution strategies have also been proposed. However one drawback of using evolution strategies for this problem can be the cost of evaluations due to the stochastic nature of the fitness function. This paper describes the use of racing algorithms to reduce the amount of evaluations of the fitness function in order to reduce the learning time. Different experiments illustrate the benefits and the limitation of racing in evolution strategies for this problem. Among the benefits is designing artificial players at the level of the top ranked players at a third of the cost.

1 Introduction

Designing artificial game players has been, and still is, studied extensively in the artificial intelligence community. This is due to games being generally interesting problems that provide several challenges (difficult or large search spaces, important branching factors, randomness etc.) Many conferences organize special tracks every year where game playing algorithms are put in competition.

Among these games, Tetris [8] is a single player game where the goal is to place randomly falling pieces onto a 10×20 game board. Each completed horizontal line is cleared from the board and scores points to the player and the goal is to clear as much lines as possible before the board is filled.

Among the challenges in learning Tetris is the prohibitive size of the search space that approaches 10^{60}, and which can only be tackled using approximations. Interestingly enough one cannot play Tetris for ever, the game ends with probability one [7]. Furthermore, it has been shown that the problem of finding strategies to maximize the score in Tetris is NP-Complete [6].

Many authors proposed algorithms to design artificial Tetris player (see below), however to this day evolutionary algorithms outperform all other methods by far. And among these the Covariance Matrix Adaptation Evolution Strategy (CMA-ES) [10] and the noisy cross entropy[3] hold the record (several million lines on average).

In the present paper we will describe the use of racing procedures in the case of CMA-ES on the Tetris learning problem in order to reduce the learning time

J.-K. Hao et al. (Eds.): EA 2011, LNCS 7401, pp. 193–204, 2012.
© Springer-Verlag Berlin Heidelberg 2012

which as it will be clear shortly, can be very problematic. We will begin with a description of the problem of learning Tetris strategies and review some of the existing work. We present the algorithm with the racing methods used, and then present some experiments. Finally we will discuss the results and give some conclusions.

1.1 Learning Tetris Strategies

In the literature, artificial Tetris players use evaluation functions that evaluate the game state by assigning numerical values. The agent will decide which action to take based on the values of this function.

When a piece is to be placed, the agent simulates and evaluates all the possible boards that would result from all the possible moves of the piece, and then chooses the move that leads to the best valued game board. Note that since the agent follows a greedy strategy, the evaluation function is the only component that enters into the decision making process. Hence, designing a Tetris player amounts to designing an evaluation function. Such a function should synthesize the game state giving high values to "good boards" and low values to "bad" ones. The evaluation function for the game of Tetris may be considered as mimicking the evaluation a human player makes when deciding where to drop the piece and which orientation to give to it. It should rank higher game states that increase the chance to clear lines in subsequent moves. For example, such boards should be low in height and should contain as less holes as possible. These properties (the height, the number of holes) are called *features* and an evaluation function is a *weighted linear combination of these features*.

Let f_i denotes a feature function, f_i maps a state (a board configuration) to a real value. The value of a state s is then defined as:

$$V(s) = \sum_{i=1}^{N} w_i f_i(s) \tag{1}$$

where N is the number of features and w_i are weights which, without loss of generality can sum to one.

Let P be the set of all pieces and D the set of all possible decisions. We note $d_i(p) \in D$ the ith decision applied on piece $p \in P$. A decision is a rotation (4 possibilities) and a translation (10 possibilities[1]) for the current piece. Furthermore, we note γ the function that maps a pair $(s, d_i(p))$, a state s and a decision $d_i(p)$, to a new state of the game after taking decision d_i for the piece p. Given all possible decisions $d_i(p) \in D$ for a current piece p in state s, the player will choose the highest valued one:

$$\hat{d}(s, p) = \arg \max_{d_i(p) \in D} V(\gamma(s, d_i(p))).$$

The function $\hat{d}(s, p)$ is the player's decision function. As stated above the only component that enter into the decision process is the value V, in other words

---

[1] The number of translations may be different depending on the game width.

the behavior of the player is conditioned by the value of the feature functions $f_{1...N}$. For a comprehensive review of the existing features the reader can see [19]. Learning tetris strategies amounts at (1) choosing a set of feature functions and (2) fixing their weights in eq. 1. In the present work we choose the set of eight features used in [19] and [5].

There are many papers that address the problem of learning Tetris strategies using different methods ranging from reinforcement learning to evolutionary computation. Dellacherie [8] fixed the weights by hand proposing good performing players. Different authors have used reinforcement learning some of which are : feature based value iteration [21], λ-policy iteration [2] and LS-λ-policy iteration [20], and approximate dynamic programming [9].

Other authors considered the problem differently and proposed to fix the feature weights using optimization techniques. The noisy cross entropy has been used by [18,19]. Evolutionary algorithms have also been used with some success. For instance [16] proposed to use genetic programming. [4] also used an evolutionary algorithm and different combinations of value function. Finally [5] proposed to use CMA-ES to optimize the weights.

At this stage the best performing player are the one proposed by [19] and [5]. They both score on average around 35 millions lines.

In all these studies, authors agree on the challenges that the problem of learning Tetris strategies introduces. First of all, the score distribution of a given strategy on a series of games follows a long tailed distribution[2] which requires to evaluate (compute the fitness) the performance of the players on a series of games rather that just one. Secondly, as the artificial player learns during the run of an algorithm and thus get better, the games it plays last longer and lengthen the evaluation time. The number of games that is required for an accurate assessment of the performance is usually a parameter of the algorithm. It is fixed empirically to obtain an enough sample to compute statistics of the performance and in the same time keep the evaluation time reasonable[3].

Different methods have been proposed to reduce the learning time in Tetris. On the one hand one can train a player on small instances of the game, in this case the possibilities to place the pieces are reduced and the player looses rapidly. On the other hand, one can increase the frequency of certain (hard to place) pieces. Here again the player looses more quickly since these pieces will disturb the games board. All these methods do not take into account the number of games played in order to estimate the score of the player which is usually a fixed number. In what follows we will describe the use of racing procedures to dynamically adapt the number of required games to evaluate the players.

1.2   Racing for Stochastic Optimization

On stochastic objective functions, the fitness values of each individual follow a distribution $\mathcal{D}$ usually unknown. We have thus $f(x) \sim \mathcal{D}_x\left(\mu_x, \sigma_x^2\right)$, where μ_x, σ_x^2

---

[2] Some authors argue that this distribution is exponential [8].

[3] Many authors reported weeks length and some times months of learning time.

are respectively the mean and the variance of the $\mathcal{D}_x$. On such problems, one cannot rely on a single evaluation of the offspring which is usually not a reliable measure, but needs to make several ones and compute statistics to estimate the true fitness.

In algorithms that rely only on the ranking of the population as it is the case for CMA-ES [10], ranking based on fitness values is not appropriate if the variability of the values is not taken into account. In such cases the ranking may not be correct and may lead to badly selected parents.

Different methods were proposed to overcome this problem. For example [11] proposes UH-CMA-ES which, among other things, increases the population variance when a change in the ranking of the population is detected. [17,15] proposed methods to reduce the number evaluations applicable for certain types distribution $\mathcal{D}$. Another way to overcome this problem is to take into account confidence bounds around the estimated fitness value using racing methods [13,1]. These methods adapts the number of evaluations dynamically until the algorithm reaches a reliable ranking in the population. [12] proposed to use such methods with CMA-ES and reported some success on different machine learning problems.

Basically these methods reevaluate only individuals where confidence intervals (CI) overlap. Reevaluation in this case is used to reduce the CI. Once enough individuals with non overlapping CI are found, the procedure stops. When the distribution $\mathcal{D}$ is unknown, the confidence intervals are computed using empirical methods using for example Hoeffding or Bernstein bounds which will be described below.

2   Learning Tetris with CMA-ES and Racing

We will begin this section by describing the CMA-ES algorithm and the racing method after which we describe their use in learning Tetris strategies. The covariance matrix adaptations evolution strategy [10] is an evolutionary algorithms that operates in continuous search spaces where the objective function f can be formulated as follows:

$$\hat{\mathbf{x}} = \arg\max_{\mathbf{x}} f(\mathbf{x}) \text{ with } f : \mathbb{R}^n \to \mathbb{R}$$

It can be seen as continuously updating a search point $m \in \mathbb{R}^n$ that is the centroid of a normally distributed population. The progress of the algorithm controls how the search distribution is updated to help the convergence to the optimum.

Alg. 1 describes the general procedure in CMA-ES. At each iteration, the algorithm samples λ points (the offspring) from the current distribution (line 5) : $\mathbf{x}_i^{t+1} \sim \mathbf{m}^t + \sigma^t \mathbf{z}_i(t)$ and $\mathbf{z}_i^t \sim \mathcal{N}(0, \mathbf{C}^t)$ $(i = 1 \cdots \lambda)$, where $\mathbf{C}^t$ is the covariance matrix and $\sigma > 0$ is a global step-size. The $\mu \leq \lambda$ best of the sampled points (the parents) are recombined by a weighted average (line 8) where $w_1 \geq \cdots \geq w_\mu > 0$ and $\sum_{i=0}^{\mu} w_i = 1$ to produce the new centroid $\mathbf{m}^{t+1}$.

Algorithm 1. Procedure $(\mu/\mu, \lambda)$-CMA-ES

Input $(\lambda, \mathbf{m}, \sigma, f, n)$

1: $\mu := \lfloor \lambda/2 \rfloor$, $\mathbf{m}^0 := \mathbf{m}$, $\sigma^0 := \sigma$, $\mathbf{C}^0 := \mathbf{I}$, $t := 1$, $\mathbf{p}_c^0 = \mathbf{p}_\sigma^0 = 0$

2: **repeat**

3:     **for** $i := 1$ **to** λ **do**

4:         $\mathbf{z}_i^t \sim \mathcal{N}_i \left(0, \mathbf{C}^t\right)$ /* Sample and evaluate offspring */

5:         $\mathbf{x}_i := \mathbf{m}^t + \sigma^t \times \mathbf{z}_i^t$

6:     **end for**

7:     $\text{race}(\mathbf{x}_1 \ldots \mathbf{x}_\lambda)$ /* Race the offspring */

8:     $\mathbf{m}^{t+1} := \sum_{i=1}^{\mu} w_i \mathbf{x}_{i:\lambda}$ /* Recombine μ best offspring, $f(\mathbf{x}_{1:\lambda}) \leq \ldots \leq f(\mathbf{x}_{\mu:\lambda})$ */

9:     $\mathbf{p}_\sigma^{t+1} := (1 - c_\sigma) \, \mathbf{p}_\sigma^t + \sqrt{c_\sigma \left(2 - c_\sigma\right) \mu_{eff}} \left(\mathbf{C}^t\right)^{\frac{1}{2}} \frac{\mathbf{m}^{t+1} - \mathbf{m}^t}{\sigma^t}$ /* Update σ */

10:     $\sigma^{t+1} := \sigma^t \exp \left(\frac{c_\sigma}{d_\sigma} \left(\frac{\|\mathbf{p}_\sigma^{t+1}\|}{\mathbb{E}[\|\mathcal{N}(0,\mathbf{I})\|]} - 1 \right) \right)$

11:     $\mathbf{p}_c^{t+1} := (1 - c_c)\mathbf{p}_c^t + \sqrt{c_c \left(2 - c_c\right) \mu_{co}} \frac{\mathbf{m}^{t+1} - \mathbf{m}^t}{\sigma^t}$ /* Update $\mathbf{C}$ */

12:     $\mathbf{C}^{t+1} := (1 - c_{co}) \, \mathbf{C}^t + \frac{c_{co}}{\mu_{co}} \mathbf{p}_c^{t+1} \left(\mathbf{p}_c^{t+1}\right)^T + c_{co} \left(1 - \frac{1}{\mu_{co}}\right) \sum_{i=1}^{\mu} w_i \left(\mathbf{z}_{i:\lambda}^t\right) \left(\mathbf{z}_{i:\lambda}^t\right)^T$

13:     $t := t + 1$

14: **until** stopping_criterion

15: **return** $\mathbf{m}^t$

Adaptation of the search distribution in CMA-ES takes place in two steps, first updating the mutation step and then updating the covariance matrix. The step-size adaptation is performed using cumulative step-size adaptation [14], where the evolution path accumulates an exponentially fading path[4] of the mean m^t (line 9) where the backward time horizon is determined by c_σ^{-1}. The adaptation of the covariance matrix C^t takes into account the change of the mean (rank-1 update), and the successful variations in the last generation (rank-μ update) (respectively lines 11 and 12). An in depth discussion on setting the parameters c_σ, d_σ, c_c, μ_{eff}, μ_{co}, c_{co} and their default values can be found in [11].

2.1   The Racing Procedure

The racing procedure we used is inspired from [12]. At each iteration of alg. 1, the λ offspring undergo multiple evaluations until either : 1) μ outstanding individuals get out of the race, in which case we are sure (with probability $(1 - \delta)$) that they are distinct. Or 2) the number of evaluations reaches some upper limit r_{limit}. Furthermore, if the actual number of evaluations necessary to distinguish μ individuals is less than the limit r_{limit}, this one is reduced:

$$r_{limit} := \max(1/\alpha \, r_{limit}, 3) \tag{2}$$

with $\alpha > 1$, and 3 is the minimum number of evaluations. On the other hand if r_{limit} evaluations where not enough to distinguish μ individuals, then r_{limit} is increased:

$$r_{limit} := \min(\alpha \, r_{limit}, r_{max}) \tag{3}$$

---

[4] The comparison of the travelled path with a path under random selection is used to adjust the step-size.

where r_{max} is a maximum number of evaluations. Initially $r_{limit} = 3$ and it is adapted at each iteration of alg. 1 using the above two rules.

After r re-evaluations, the empirical bounds $c_{i,r}$ around the mean fitness of each individual $\hat{X}_{i,r} = \frac{1}{r}\sum_{j=1}^{r} f^j(x_i)$ where $i = 1\ldots\lambda$, are computed with:

$$c_{i,r}^h = R\sqrt{\frac{\log(2n_b) - \log(\delta)}{2r}} \qquad (4)$$

using the Hoeffding bound and

$$c_{i,r}^b = \hat{\sigma}_{i,r}\sqrt{2\frac{\log(3n_b) - \log(\delta)}{r}} + 3R\frac{\log(3n_b) - \log(\delta)}{r} \qquad (5)$$

using Bernstein. Where $n_b \leq \lambda r_{limit}$ is the number of evaluations in the current race, $\hat{\sigma}_{i,r}^2 = \frac{1}{r}\sum_{j=1}^{r}\left(f^j(x_i) - \hat{X}_{i,r}\right)^2$ is the standard deviation of the fitness for individual x_i, and $(1 - \delta)$ is the level of confidence we fix. $R = |a - b|$ such that the fitness values of the offspring are almost surely between a and b. These two constant are problem dependent.

After each evaluation in the race the lower bounds $lb_{i,r}$ and the upper bounds $ub_{i,r}$ around the mean of each offspring are updated to the tightest values: $lb_{i,r} = max\left(lb_{i,r-1}, \hat{X}_{i,r} - c_{i,r}^{h/b}\right)$ and $ub_{i,r} = min\left(lb_{i,r-1}, \hat{X}_{i,r} + c_{i,r}^{h/b}\right)$

Beside the above empirical bounds, there exists for the game of Tetris estimated bounds on the mean score of the player [19] which we also used in the racing procedure. We have:

$$|X - \hat{X}|/\hat{X} \leq 2/\sqrt{n} \qquad (6)$$

with probability 0.95, where $\hat{X}$ is the average score of the player on n games and X is the expected score. In the remainder we will refer to this bound as the Tetris bound.

3   Experiments

In the following, we present few experiments we conducted in different settings to compare the effect of the racing procedures described above. In all experiments the initial search point x_0 was drawn randomly with $\|x_0\| = 1$ and the initial step-size $\sigma_0 = 0.5$. We let the algorithm run for 25 iterations (improvements for larger values were not noticeable), we set $r_{max} = 100$ and $\delta = 0.05$. There are eight feature functions in the evaluation function therefore the dimension of the search space is $n = 8$. CMA-ES parameters c_σ, d_σ, c_c, μ_{eff}, μ_{co} and c_{co} were set to their default values as presented in [11].

In order to reduce the learning time, we adopt the same scheme as in [5] and evaluate the offspring on harder games, in which "S" and "Z" appear four times more frequently than in the standard game. Experiments showed that learning on this setting does not impair the performance of the player on the standard game.

The algorithm maximizes the score of the game i.e. the fitness function is the average of lines scored by a search point on a number of games. When racing is applied, this number of games can go from 3 to r_{max}. When it is not applied it is r_{max}. The initial bounds a and b used to compute R (eq. 5) were fixed experimentally to $a = 150$ and $b = 2000$.

The games were simulated on the MDPTeris platform[5]. All the expriments were repeated 50 times and the curves represent median values.

An important issue (incidentally ignored by many authors) about learning Tetris players and in general about learning the weights of evaluation functions for games, is that the weight should be normalized. This fact is described in [5] where empirical evidence show how in CMA-ES the step-size and the principal axes of the covariance matrix diverge. In these experiments we follow the same steps and normalize the offspring after the sampling step.

Furthermore, in order to provide a coherent ranking of the offspring, we evaluate them on the same series of games. At each iteration, a series of r_{max} independent games (random seeds) are drawn randomly and each individual of the λ offspring plays the same games. This way we can assess of the offspring's performance on the same setting.

3.1   Results and Discussions

The effect of racing is clear on the learning time of the algorithm. The number of evaluations is indeed reduced without a loss of performance. Fig. 1 on the right shows the fitness function of the mean vector (the centroid of the distribution) for the different racing procedures and without racing. All algorithms reach the same performance at different evaluations costs. Hoeffding bounds perform the best reaching the performance for the smallest evaluation time, followed by Tetris bounds and Bernstein. Fig. 1 on the left shows the median rate of evaluations (r_{limit} being 100%) used in the races. With Tetris bounds this number of evaluation is reduced at the beginning of the learning process and increases towards the end reaching r_{limit}. For Hoeffding bounds the number of evaluations oscillates between r_{limit} and lower values. Finally, Bernstein bounds were not efficient compared to Hoeffding bounds, the number of evaluations is not reduced and at each race each individual is evaluated r_{limit} times. This is due to the fact that the standard deviation of the fitness is of the same order of its average. This is the case for Tetris, the standard deviation of the score distribution is almost equal to its mean. In other words $\hat{\sigma}_{i,t}$ is large compared to R in eq. 5. When this is the case Hoeffding bounds are tighter than Bernstein's, and this is the case in all our experiments.

Recall that r_{limit} is adapted using equations 2 and 3 and is initially set to 3, which explains why in the right of fig. 1, Bernstein races used less evaluations than in the case without racing.

We noticed also that there is a large variability in the racing procedure. Fig. 2 shows the rate of evaluation out of r_{limit} of all 50 runs for Hoeffding and Tetris bounds.

---

[5] MDPTetris simulator is available at http://mdptetris.gforge.inria.fr

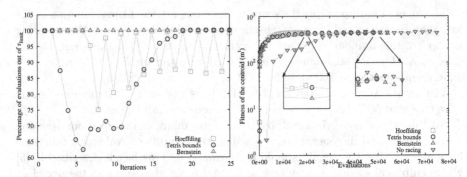

Fig. 1. On the left, the number of evaluations with racing using different bounds versus iterations. On the right, log of the fitness of the mean vector m^t versus the number of evaluations. Median values over 50 runs.

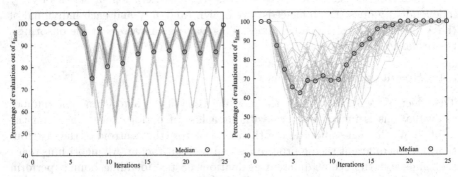

Fig. 2. The number of evaluations for 50 runs thick line represent the median value. With Hoeffding races (left). Races with Tetris bounds(right).

It is interesting to notice that the racing procedures reduce the number of evaluations early in the process and do not perform as well at the end. At this stage the racing procedure cannot distinguish between the offspring and selection is made only based on the fitness values (averaged over the r_{limit} reevaluation). The CI around the offspring's fitness overlap indicating that they perform equally well. This also suggest (see below) that the ranking of the offspring might not be correct. Incidentally, this correspond to the time where the convergence of step-sizes start to "flatten" (figures 4 and 5).

The effect of the population size on the racing procedure can be seen on fig. 3 where is shown the number of evaluations out of r_{limit} for different population sizes. Apparently increasing the population size reduces the number of evaluations within the race using Hoeffding bounds. On the other hand the opposite can be seen when using Tetris bounds.

Furthermore the population size does not effect the performance of the algorithms, the median value of the mean vector fitness reach similar values for different population sizes. This is the case for both Tetris and Hoeffding bounds (figures 4 and 5).

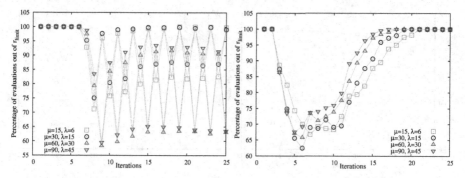

Fig. 3. The median number of evaluations over 50 runs for different population sizes. Using Hoeffding races (left). Using Tetris bounds (right)

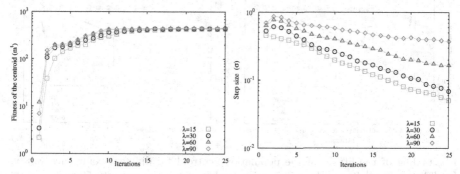

Fig. 4. CMA-ES and Tetris bounds with multiple population sizes. Log of the fitness of the mean vector m^t (left) and the step-size σ (right). Median values over 50 runs.

Fig. 6 shows the effect of the confidence value on the evaluation time. It presents four setting with different confidence $(1 - \delta)$ levels. Using Hoeffding bounds, if we reduce the confidence value the algorithm stops earlier. This is the expected behavior: reducing the confidence reduces the bounds and races are decided faster. On the other hand, Bernstein bounds are not affected at all. The races are not able to distinguish statistical differences between the offspring even with low confidence values.

In order to assess the results of the racing procedure, we test the learned players on a standard game[6] and compare the scores with previously published scores. We follow the same scheme as in [5] to construct the player. We record the last mean vector m^t of each run of the algorithm and compute the mean of all these vectors (in our setting 50). This is merely a convention and others are possible[7]. The score of a player is the average score on 100 games. Taking into

---

[6] A game of size 10×20 with all seven pieces equally probable.

[7] One could also choose the best performing vector out of the whole set. However testing them all on a reasonable number of games might be long. For example testing one player on 10 games took 15 hours on a 2 Ghz CPU.

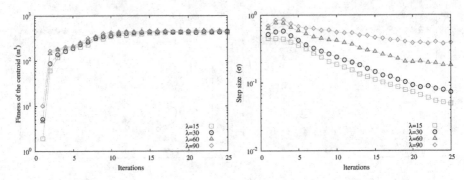

Fig. 5. CMA-ES and Hoeffding races with multiple population sizes. Log of the fitness of the mean vector m^t (left) and the step-size σ (right). Median values over 50 runs.

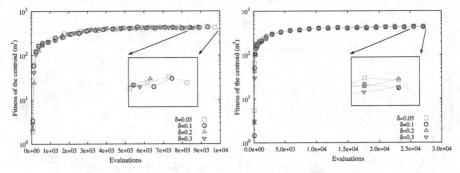

Fig. 6. Log of the fitness of the population centroid for different confidence values. Hoeffding races (left) and Bernstein races (right). Median values over 50 runs.

Table 1. Comparison of the scores of the learned strategies on two game sizes

Strategy	10×16	10×20	Evaluations
No racing [5]	8.7×10^5	36.3×10^6	100%
Tetris bounds	8.1×10^5	31.5×10^6	67%
Hoeffding	7.8×10^5	33.2×10^6	27%

account variability of the score distribution, a confidence bound of $\pm 20\%$ is to be taken into account (see eq.6).

Table 1 present the scores of the players learned on CMA-ES with racing compared to the score reported by [5]. We notice that for both games instances the scores are statistically equivalent. Therefore we can conclude that the racing procedures do not affect the performance of the player. Furthermore, the same performance was obtained using, in the case of Hoeffding bounds two thirds less evaluations and in the case of Tetris bounds one third less evaluations.

3.2 Further Thoughts

Even though there are benefits in using racing, there are also some disadvantages. First of all, the choice of a value for r_{max}, is crucial for time consuming problems.

A too low value leads to short races and thus to no statistical soundness. On the other hand giving a hight value may lead to too many evaluations once the algorithm converges.

As said earlier, when the algorithm converges, the confidence intervals overlap and the racing procedure cannot distinguish statistical differences. This could be considered as a limitation in problems where the cost of an evaluation is not the same at the beginning of the evaluation and towards the end, as it is the case for Tetris. If given the choice, one would prefer to have less costly evaluations rather than the converse. In problems where the cost of the evaluation is constant, this issue is not as severe.

Furthermore, in the event the offspring converge, they all have the same fitness (again statistically). This could raise problems in algorithms that select based on ranking. In the experiments we performed, we noticed that in some cases the ranking (based only on the fitness values) of the offspring changes at each reevaluation during the race. This indicates that at no time the ranking of the population is correct. Selection in such cases becomes random which, and this is purely speculative, might explain why the step-size increases at that stage (figures 4 and 5).

One way to circumvent this limitation, would be to use this convergence as an indication to restart procedure. Once the population converges and we cannot distinguish different individuals, start the algorithm over. It could also be used as a stopping criterion.

4   Conclusions

In the present work, we have described the use of racing methods to reduce the evaluation time in learning artificial Tetris players. Designing such players was done in the past by performing the learning on reduced instances of the game. The addition of racing methods can reduce significantly the learning time without loss of performance, as it has been shown in the experiments.

These experiments also showed that the population size does not affect the racing procedure and its performance. Using Hoeffding and Tetris bounds allowed to reduce the evaluation time, on the other hand Bernstein bounds were inefficient in all our problem instances due the properties of the Tetris fitness function. This also was the case when the confidence level was lowered.

Testing on the standard game instances allowed to show that players designed using CMA-ES with racing have the same performance as the best existing players. Racing also raises few questions that we leave for further investigations.

References

1. Audibert, J.-Y., Munos, R., Szepesvári, C.: Tuning Bandit Algorithms in Stochastic Environments. In: Hutter, M., Servedio, R.A., Takimoto, E. (eds.) ALT 2007. LNCS (LNAI), vol. 4754, pp. 150–165. Springer, Heidelberg (2007)
2. Bertsekas, D., Tsitsiklis, J.: Neuro-Dynamic Programming. Athena Scientific (1996)

3. de Boer, P., Kroese, D., Mannor, S., Rubinstein, R.: A tutorial on the cross-entropy method. Annals of Operations Research 1(134), 19–67 (2004)
4. Böhm, N., Kókai, G., Mandl, S.: An Evolutionary Approach to Tetris. In: University of Vienna Faculty of Business; Economics, Statistics (eds.) Proc. of the 6th Metaheuristics International Conference, CDROM (2005)
5. Boumaza, A.: On the evolution of artificial tetris players. In: Proc. of the IEEE Symp. on Comp. Intel. and Games, CIG 2009, pp. 387–393. IEEE (June 2009)
6. Burgiel, H.: How to lose at Tetris. Mathematical Gazette 81, 194–200 (1997)
7. Demaine, E.D., Hohenberger, S., Liben-Nowell, D.: Tetris is Hard, Even to Approximate. In: Warnow, T.J., Zhu, B. (eds.) COCOON 2003. LNCS, vol. 2697, pp. 351–363. Springer, Heidelberg (2003)
8. Fahey, C.P.: Tetris AI, Computer plays Tetris (2003), on the web http://colinfahey.com/tetris/tetris_en.html
9. Farias, V., van Roy, B.: Tetris: A study of randomized constraint sampling. Springer (2006)
10. Hansen, N., Müller, S., Koumoutsakos, P.: Reducing the time complexity of the derandomized evolution strategy with covariance matrix adaptation (CMA-ES). Evolutionary Computation 11(1), 1–18 (2003)
11. Hansen, N., Niederberger, S., Guzzella, L., Koumoutsakos, P.: A method for handling uncertainty in evolutionary optimization with an application to feedback control of combustion. IEEE Trans. Evol. Comp. 13(1), 180–197 (2009)
12. Heidrich-Meisner, V., Igel, C.: Hoeffding and bernstein races for selecting policies in evolutionary direct policy search. In: Proc. of the 26th ICML, pp. 401–408. ACM, New York (2009)
13. Maron, O., Moore, A.W.: Hoeffding races: Accelerating model selection search for classification and function approximation. In: Proc. Advances in Neural Information Processing Systems, pp. 59–66. Morgan Kaufmann (1994)
14. Ostermeier, A., Gawelczyk, A., Hansen, N.: A derandomized approach to self-adaptation of evolution strategies. Evolutionary Computation 2(4), 369–380 (1994)
15. Schmidt, C., Branke, J., Chick, S.E.: Integrating Techniques from Statistical Ranking into Evolutionary Algorithms. In: Rothlauf, F., Branke, J., Cagnoni, S., Costa, E., Cotta, C., Drechsler, R., Lutton, E., Machado, P., Moore, J.H., Romero, J., Smith, G.D., Squillero, G., Takagi, H. (eds.) EvoWorkshops 2006. LNCS, vol. 3907, pp. 752–763. Springer, Heidelberg (2006)
16. Siegel, E.V., Chaffee, A.D.: Genetically optimizing the speed of programs evolved to play tetris. In: Angeline, P.J., Kinnear Jr., K.E. (eds.) Advances in Genetic Programming 2, pp. 279–298. MIT Press, Cambridge (1996)
17. Stagge, P.: Averaging Efficiently in the Presence of Noise. In: Eiben, A.E., Bäck, T., Schoenauer, M., Schwefel, H.-P. (eds.) PPSN 1998. LNCS, vol. 1498, pp. 188–197. Springer, Heidelberg (1998)
18. Szita, I., Lörincz, A.: Learning tetris using the noisy cross-entropy method. Neural Comput. 18(12), 2936–2941 (2006)
19. Thiery, C., Scherrer, B.: Building Controllers for Tetris. International Computer Games Association Journal 32, 3–11 (2009)
20. Thiery, C., Scherrer, B.: Least-Squares λ Policy Iteration: Bias-Variance Trade-off in Control Problems. In: Proc. ICML, Haifa (2010)
21. Tsitsiklis, J.N., van Roy, B.: Feature-based methods for large scale dynamic programming. Machine Learning 22, 59–94 (1996)

Black-Box Complexity: Breaking the $O(n \log n)$ Barrier of LeadingOnes

Benjamin Doerr and Carola Winzen

Max-Planck-Institut für Informatik, Saarbrücken, Germany

Abstract. We show that the unrestricted black-box complexity of the n-dimensional XOR- and permutation-invariant LeadingOnes function class is $O(n \log(n)/ \log \log n)$. This shows that the recent natural looking $O(n \log n)$ bound is not tight.

The black-box optimization algorithm leading to this bound can be implemented in a way that only 3-ary unbiased variation operators are used. Hence our bound is also valid for the unbiased black-box complexity recently introduced by Lehre and Witt. The bound also remains valid if we impose the additional restriction that the black-box algorithm does not have access to the objective values but only to their relative order (ranking-based black-box complexity).

Keywords: Algorithms, black-box complexity, query complexity, run-time analysis, theory.

1 Introduction

The black-box complexity of a set $\mathcal{F}$ of functions $\mathcal{S} \to \mathbb{R}$, roughly speaking, is the number of function evaluations necessary to find the maximum of any member of $\mathcal{F}$ which—apart from the points evaluated so far—is unknown. This and related notions are used to describe how difficult a problem is to be solved via general-purpose (randomized) search heuristics. Consequently, black-box complexities are very general lower bounds which are valid for a wide range of evolutionary algorithms. A number of different black-box notions exist, each capturing different classes of randomized search heuristics, cf. [1–4].

In this paper we consider the unrestricted black-box model by Droste, Jansen, and Wegener [1] and the unbiased black-box model by Lehre and Witt [2], and we shortly remark on the ranking-based models which we propose in [3]. A formal definition of the first two models is given in Section 2. For now, let us just mention that algorithms in the unrestricted black-box model are allowed to query any bit string, whereas in the k-ary unbiased model, an algorithm may only query search points sampled from a k-ary unbiased distribution. That is, in each iteration, the algorithm may either query a random search point or, based upon at most k previously queried search points, it may generate a new one. The new search point can be generated only by using so-called unbiased variation operators. These are operators that are symmetric both in the bit values (0 or 1) and the

J.-K. Hao et al. (Eds.): EA 2011, LNCS 7401, pp. 205–216, 2012.

bit positions. More formally, the variation operation must be invariant under all automorphisms of the hypercube.

In this work, we are concerned with the black-box complexity of the LEADINGONES function, which is one of the classical test functions for analyzing the optimization behavior of different search heuristics. The function itself is defined via LO : $\{0,1\}^n \to [0..n], x \mapsto \max\{i \in [0..n] \mid \forall j \leq i : x_j = 1\}$. It was introduced in [5] to disprove a previous conjecture by Mühlenbein [6] that any unimodal function can be optimized by the well-known $(1+1)$ evolutionary algorithm (EA) in $O(n \log n)$ iterations. Rudolph [5] proves an upper bound of $O(n^2)$ for the expected optimization time of the $(1+1)$ EA on LO and concludes from experimental studies a lower bound of $\Omega(n^2)$—a bound which was rigorously proven in 2002 by Droste, Jansen, and Wegener [7]. This $\Theta(n^2)$ expected optimization time of the simple $(1+1)$ EA seems optimal among the commonly studied evolutionary algorithms.

Note that the unrestricted black-box complexity of the LO function is 1: The algorithm querying the all-ones vector $(1, \ldots, 1)$ in the first query is optimal. Motivated by this and by the fact that the unbiased black-box model only allows variation operators which are invariant with respect to the bit values and the bit positions, we shall study here a generalized version of the LO function. More precisely, we consider the closure of LO under all permutations $\sigma \in S_n$ and under all exchanges of the bit values 0 and 1. It is immediate, that each of these functions has a fitness landscape that is isomorphic to the one induced by LO. To be more precise, we define for any bit string $z \in \{0,1\}^n$ and any permutation σ of $[n]$ the function

$$\text{LO}_{z,\sigma} : \{0,1\}^n \to [0..n], x \mapsto \max\{i \in [0..n] \mid \forall j \leq i : z_{\sigma(j)} = x_{\sigma(j)}\}.$$

We let LEADINGONES$_n$ be the set $\{\text{LO}_{z,\sigma} \mid z \in \{0,1\}^n, \sigma \in S_n\}$ of all such functions.

Note that this definition differs from the one in [1], where only the subclass LEADINGONES$_n^0 := \{\text{LO}_{z,\text{id}} \mid z \in \{0,1\}^n\}$ not invariant under permutations of $[n]$ is studied. Here, id denotes the identity mapping on $[n]$. For this restricted subclass, Droste, Jansen, and Wegener [1] prove an unrestricted black-box complexity of $\Theta(n)$. Of course, their lower bound $\Omega(n)$ is a lower bound for the unrestricted black-box complexity of the general LEADINGONES$_n$ function class, and consequently, a lower bound for the unbiased black-box complexity of LEADINGONES$_n$.

The function class LEADINGONES$_n$ has implicitly been studied for the first time in [2], where Lehre and Witt show that indeed the $(1+1)$ EA is provably (asymptotically) optimal among all unbiased black-box algorithms of arity at most one. This establishes a natural $\Theta(n^2)$ bound for LEADINGONES$_n$.

Surprisingly, it turns out that this bound does not hold for the unrestricted black-box model and that it does not even hold in the 2-ary unbiased black-box model. In [8] it is shown that, assuming knowledge on $\sigma(1), \ldots, \sigma(\ell)$, one can perform a binary search to determine $\sigma(\ell + 1)$ and its corresponding bit value. Since this has to be done n times, an upper bound of $O(n \log n)$ for the unrestricted black-box complexity of LEADINGONES$_n$ follows. Furthermore, this $O(n \log n)$ bound can already be achieved in the binary unbiased model. Up to now, this is the best known upper bound for the unrestricted and the 2-ary unbiased black-box complexity of LEADINGONES$_n$.

In this work we show that both in the unrestricted model (Section 3) and for arities at least three (Section 4), one can do better, namely that $O(n \log(n) / \log \log n)$ queries suffice to optimize any function in LEADINGONES$_n$. This breaks the previous $O(n \log n)$ barrier. This result also shows why previous attempts to prove an $\Omega(n \log n)$ lower bound must fail.

Unfortunately, also the ranking-based model does not help to overcome this unnatural low black-box complexity. We shall comment in Section 5 that the 3-ary unbiased ranking-based black-box complexity of LEADINGONES$_n$, too, is $O(n \log(n) / \log \log n)$.

As for the memory-restricted model we note without proof that a memory of size $O(\sqrt{\log n})$ suffices to achieve the same bound.

2   Preliminaries

In this section we briefly introduce the two black-box models considered in this work, the *unrestricted black-box model* by Droste, Jansen, and Wegener [1] and the *unbiased black-box model* by Lehre and Witt [2]. Due to space limitations, we keep the presentation as concise as possible. For a more detailed discussion of the two different black-box models and for the definition of the ranking-based versions considered in Section 5, we refer the reader to [3].

Before we introduce the two black-box models, let us fix some notation. For all positive integers $k \in \mathbb{N}$ we abbreviate $[k] := \{1, \ldots, k\}$ and $[0..k] := [k] \cup \{0\}$. By e_k^n we denote the k-th unit vector $(0, \ldots, 0, 1, 0, \ldots, 0)$ of length n. For a set $I \subseteq [n]$ we abbreviate $e_I^n := \sum_{i \in I} e_i^n = \oplus_{i \in I} e_i^n$, where $\oplus$ denotes the bitwise exclusive-or. By S_n we denote the set of all permutations of $[n]$ and for $x = (x_1, \ldots, x_n) \in \{0,1\}^n$ and $\sigma \in S_n$ we abbreviate $\sigma(x) := (x_{\sigma(1)}, \ldots, x_{\sigma(n)})$. For any two strings $x, y \in \{0,1\}^n$ let $\mathcal{B}(x, y) := \{i \in [n] \mid x_i = y_i\}$, the set of positions in which x and y coincide.

For $r \in \mathbb{R}_{\geq 0}$, let $\lceil r \rceil := \min\{n \in \mathbb{N}_0 \mid n \geq r\}$ and $\lfloor r \rfloor := \max\{n \in \mathbb{N}_0 \mid n \leq r\}$. For the purpose of readability we sometime omit the $\lceil \cdot \rceil$ signs, that is, whenever we write r where an integer is required, we implicitly mean $\lceil r \rceil$. All logarithms log in this work are base two logarithms. By ln we denote the logarithm to the base $e := \exp(1)$.

Black-Box Complexity. Let $\mathcal{A}$ be a class of algorithms and $\mathcal{F}$ be a class of functions. For every $A \in \mathcal{A}$ and $f \in \mathcal{F}$ let $T(A, f) \in \mathbb{R} \cup \{\infty\}$ be the expected number of fitness evaluations until A queries for the first time some $x \in \arg \max f$. We call $T(A, f)$ the *expected optimization time* of A for f. The $\mathcal{A}$-*black-box complexity* of $\mathcal{F}$ is $T(A, \mathcal{F}) := \sup_{f \in \mathcal{F}} T(A, f)$, the worst-case expected optimization time of A on $\mathcal{F}$. The $\mathcal{A}$-*black-box complexity of* $\mathcal{F}$ is $\inf_{A \in \mathcal{A}} T(A, \mathcal{F})$, the best worst-case expected optimization time an algorithm of $\mathcal{A}$ can exhibit on $\mathcal{F}$.

```
1  Initialization:
2     Sample x^(0) according to some probability distribution p^(0) on S;
3     Query f(x^(0));
4  Optimization:
5     for t = 1, 2, 3, ... do
6        Depending on ((x^(0), f(x^(0))), ..., (x^(t-1), f(x^(t-1)))) choose a probability
          distribution p^(t) on S;
7        Sample x^(t) according to p^(t);
8        Query f(x^(t));
```

Algorithm 1. Scheme of an unrestricted black-box algorithm for optimizing $f : \mathcal{S} \to \mathbb{R}$

The Unrestricted Black-Box Model. The black-box complexity of a class of functions depends crucially on the class of algorithms under consideration. If the class $\mathcal{A}$ contains all (deterministic and randomized) algorithms, we refer to the respective complexity as the *unrestricted black-box complexity*. This is the model by Droste, Jansen, and Wegener [1]. The scheme of an unrestricted algorithm is presented in Algorithm 1.

Note that this algorithm runs forever. Since our performance measure is the expected number of iterations needed until for the first time an optimal search point is queried, we do not specify a termination criterion for black-box algorithms here.

The Unbiased Black-Box Model. As observed already by Droste, Jansen, and Wegener [1], the unrestricted black-box complexity can be surprisingly low for different function classes. For example, it is shown in [1] that the unrestricted black-box complexity of the NP-hard optimization problem MAXCLIQUE is polynomial.

This motivated Lehre and Witt [2] to define a more restrictive class of algorithms, the so-called *unbiased black-box model*, where algorithms may generate new solution candidates only from random or previously generated search points and only by using *unbiased* variation operators, cf. Definition 1. Still the model captures most of the commonly studied search heuristics, such as many $(\mu + \lambda)$ and (μ, λ) evolutionary algorithms, simulated annealing algorithms, the Metropolis algorithm, and the Randomized Local Search algorithm (confer the book [9] for the definitions of these algorithms).

Definition 1 (Unbiased Variation Operator). *For all $k \in \mathbb{N}$, a k-ary unbiased distribution $\left(D(\cdot \mid y^{(1)}, \ldots, y^{(k)})\right)_{y^{(1)}, \ldots, y^{(k)} \in \{0,1\}^n}$ is a family of probability distributions over $\{0,1\}^n$ such that for all inputs $y^{(1)}, \ldots, y^{(k)} \in \{0,1\}^n$ the following two conditions hold.*

(i) $\forall x, z \in \{0,1\}^n : D(x \mid y^{(1)}, \ldots, y^{(k)}) = D(x \oplus z \mid y^{(1)} \oplus z, \ldots, y^{(k)} \oplus z)$;

(ii) $\forall x \in \{0,1\}^n \, \forall \sigma \in S_n : D(x \mid y^{(1)}, \ldots, y^{(k)}) = D(\sigma(x) \mid \sigma(y^{(1)}), \ldots, \sigma(y^{(k)}))$.

```
1  Initialization:
2     Sample x^(0) ∈ {0,1}^n uniformly at random;
3     Query f(x^(0));
4  Optimization:
5     for t = 1, 2, 3, ... do
6        Depending on (f(x^(0)), ..., f(x^(t−1))) choose up to k indices
          i_1, ..., i_k ∈ [0..t − 1] and a k-ary unbiased distribution
          D(· | x^(i_1), ..., x^(i_k));
7        Sample x^(t) according to D(· | x^(i_1), ..., x^(i_k));
8        Query f(x^(t));
```

Algorithm 2. Scheme of a k-ary unbiased black-box algorithm

We refer to the first condition as $\oplus$-*invariance and to the second as* permutation invariance. *An operator sampling from a k-ary unbiased distribution is called a k-ary unbiased variation operator.*

1-ary—also called *unary*—operators are sometimes referred to as mutation operators, in particular in the field of evolutionary computation. 2-ary—also called *binary*—operators are often referred to as crossover operators. If we allow arbitrary arity, we call the corresponding model the *-ary unbiased black-box model.

A k-ary unbiased black-box algorithm can now be described via the scheme of Algorithm 2. The *k-ary unbiased black-box complexity* of some class of functions $\mathcal{F}$ is the complexity of $\mathcal{F}$ with respect to all k-ary unbiased black-box algorithms.

3 On LeadingOnes$_n$ in the Unrestricted Model

This section is devoted to the main contribution of this work, Theorem 1. Recall from the introduction that we have defined

$$\text{Lo}_{z,\sigma} : \{0,1\}^n \to [0..n], x \mapsto \max\{i \in [0..n] \mid \forall j \le i : z_{\sigma(j)} = x_{\sigma(j)}\}$$

and $\text{LEADINGONES}_n := \{\text{Lo}_{z,\sigma} \mid z \in \{0,1\}^n, \sigma \in S_n\}$.

Theorem 1. *The unrestricted black-box complexity of* LEADINGONES$_n$ *is* $O(n \log(n)/\log \log n)$.

The proof of Theorem 1 is technical. For this reason, we split it into several lemmata. The main proof can be found at the end of this section. We remark already here that the algorithm certifying Theorem 1 will make use of unbiased variation operators only. Hence, it also proves that the *-ary unbiased black-box complexity of LEADINGONES$_n$ is $O(n \log(n)/\log \log n)$. This will be improved to the 3-ary model in Section 4.

The main idea of both the *-ary and the 3-ary algorithm is the following. Given a bit string x of fitness $\text{Lo}_{z,\sigma}(x) = \ell$, we iteratively first learn $k := \sqrt{\log n}$ bit positions $\sigma(\ell + 1), ..., \sigma(\ell + k)$ and their corresponding bit values which we fix for all further iterations of the algorithm. Learning such a block of size k

will require $O(k^3/\log k^2)$ queries. Since we have to optimize n/k such blocks, the overall expected optimization is $O(nk^2/\log k^2) = O(n \log(n)/\log\log n)$. In what follows, we shall formalize this idea.

Convention: For all following statements let us fix a positive integer n, a bit string $z \in \{0,1\}^n$ and a permutation $\sigma \in S_n$.

Definition 2 (Encoding Pairs). *Let $\ell \in [0..n]$ and let $y \in \{0,1\}^n$ with $\mathrm{LO}_{z,\sigma}(y) = \ell$. If $x \in \{0,1\}^n$ satisfies $\mathrm{LO}_{z,\sigma}(x) \geq \mathrm{LO}_{z,\sigma}(y)$ and $\ell = |\{i \in [n] \mid x_i = y_i\}|$, we call (x,y) an ℓ-encoding pair for $\mathrm{LO}_{z,\sigma}$.*

If (x,y) is an ℓ-encoding pair for $\mathrm{LO}_{z,\sigma}$, the bit positions $\mathcal{B}(x,y)$ are called the ℓ-encoding bit positions of $\mathrm{LO}_{z,\sigma}$ and the bit positions $j \in [n]\backslash\mathcal{B}(x,y)$ are called non-encoding.

If (x,y) is an ℓ-encoding pair for $\mathrm{LO}_{z,\sigma}$ we clearly either have have $\mathrm{LO}_{z,\sigma}(x) > \ell$ or $\mathrm{LO}_{z,\sigma}(x) = \mathrm{LO}_{z,\sigma}(y) = n$. For each non-optimal $y \in \{0,1\}^n$ we call the unique bit position which needs to be flipped in y in order to increase the objective value of y the ℓ-critical bit position of $\mathrm{LO}_{z,\sigma}$. Clearly, the ℓ-critical bit position of $\mathrm{LO}_{z,\sigma}$ equals $\sigma(\ell+1)$, but since σ is unknown to the algorithm we shall make use of this knowledge only in the analysis, and not in the definition of our algorithms. In the same spirit, we call the k bit positions $\sigma(\ell+1),\ldots,\sigma(\ell+k)$ the k ℓ-critical bit positions of $\mathrm{LO}_{z,\sigma}$.

Lemma 1. *Let $\ell \in [0..n-1]$ and let $(x,y) \in \{0,1\}^n \times \{0,1\}^n$ be an ℓ-encoding pair for $\mathrm{LO}_{z,\sigma}$. Furthermore, let $k \in [n - \mathrm{LO}_{z,\sigma}(y)]$ and let $y' \in \{0,1\}^n$ with $\ell \leq \mathrm{LO}_{z,\sigma}(y') < \ell + k$.*

If we create y'' from y' by flipping each non-encoding bit position $j \in [n]\backslash\mathcal{B}(x,y)$ with probability $1/k$, then

$$\Pr[\mathrm{LO}_{z,\sigma}(y'') > \mathrm{LO}_{z,\sigma}(y')] \geq (ek)^{-1}.$$

Lemma 1 motivates us to formulate Algorithm 3 which can be seen as a variant of the standard $(1+1)$ EA, in which we fix some bit positions and where we apply a non-standard mutation probability. The variation operator $\mathtt{random}(y',x,y,1/k)$ samples a bit string y'' from y' by flipping each non-encoding bit position $j \in [n]\backslash\mathcal{B}(x,y)$ with probability $1/k$. This is easily seen to be an unbiased variation operator of arity 3.

The following statement follows easily from Lemma 1 and the linearity of expectation.

Corollary 1. *Let (x,y), ℓ, and k be as in Lemma 1. Then the (x,y)-encoded $(1+1)$ EA with mutation probability $1/k$, after an expected number of $O(k^2)$ queries, outputs a bit string $y' \in \{0,1\}^n$ with $\mathrm{LO}_{z,\sigma}(y') \geq \ell + k$.*

A second key argument in the proof of Theorem 1 is the following. Given an ℓ-encoding pair (x,y) and a bit string y' with $\mathrm{LO}_{z,\sigma}(y') \geq \ell + \sqrt{\log n}$, we are able to learn the $\sqrt{\log n}$ ℓ-critical bit positions $\sigma(\ell+1),\ldots,\sigma(\ell+\sqrt{\log n})$ in an expected number of $O(\log^{3/2}(n)/\log\log n)$ queries. This will be formalized in the following statements.

1 **Input:** ℓ-encoding pair $(x, y) \in \{0, 1\}^n \times \{0, 1\}^n$.
2 $y' \leftarrow y$;
3 **while** $\mathrm{Lo}_{z,\sigma}(y') < \mathrm{Lo}_{z,\sigma}(y) + k$ **do**
4      $y'' \leftarrow \mathtt{random}(y', x, y, 1/k)$;
5   Query $\mathrm{Lo}_{z,\sigma}(y'')$;
6   **if** $\mathrm{Lo}_{z,\sigma}(y'') > \mathrm{Lo}_{z,\sigma}(y')$ **then** $y' \leftarrow y''$;
7 **Output** y';

Algorithm 3. The (x, y)-encoded $(1+1)$ evolutionary algorithm with mutation probability $1/k$

Lemma 2. *Let* $\ell \in [0..n - \lceil\sqrt{\log n}\rceil]$ *and let* (x, y) *be an* ℓ-*encoding pair for* $\mathrm{Lo}_{z,\sigma}$. *Furthermore, let* y' *be a bit string with* $\mathrm{Lo}_{z,\sigma}(y') \geq \ell + \sqrt{\log n}$.

For each $i \in [8e \log^{3/2}(n)/\log\log n]$ *let* y^i *be sampled from* y' *by independently flipping each non-encoding bit position* $j \in [n]\backslash\mathcal{B}(x, y)$ *with probability* $1/\sqrt{\log n}$.
For each $c \in [\sqrt{\log n}]$ *let*

$$X_c := \{y^i \mid i \in [8e \log^{3/2}(n)/\log\log n] \text{ and } \mathrm{Lo}_{z,\sigma}(y^i) = \ell + c - 1\},$$

the set of all samples y^i *with* $\mathrm{Lo}_{z,\sigma}(y^i) = \ell + c - 1$.
Then

$$\Pr\left[\forall c \in [\sqrt{\log n}] : |X_c| \geq 4\log(n)/\log\log n\right] \geq 1 - o(1).$$

These sets X_c are large enough to identify $\sigma(\ell + c)$.

Lemma 3. *Let* ℓ, (x, y), *and* y' *be as in Lemma 2 and let* $t := 4\log(n)/\log\log n$.

For any $c \in [\sqrt{\log n}]$ *let* X_c *be a set of at least* t *bit strings* $y^1(c), \ldots, y^{|X_c|}(c)$ *with fitness* $\mathrm{Lo}_{z,\sigma}(y^i(c)) = \ell + c - 1$, *which are sampled from* y' *by independently flipping each non-encoding bit position* $j \in [n]\backslash\mathcal{B}(x, y)$ *with probability* $1/\sqrt{\log n}$.

Then we have, with probability at least $1 - o(1)$, *that for all* $c \in [\sqrt{\log n}]$ *there exists only one non-encoding* $j := j_{\ell+c} \in [n]\backslash\mathcal{B}(x, y)$ *with* $y'_j = 1 - y^i_j(c)$ *for all* $i \in [|X_c|]$. *Clearly,* $j = \sigma(\ell + c)$.

Combining Lemma 2 with Lemma 3 we immediately gain the following.

Corollary 2. *Let* $\ell, (x, y), y'$, *and* $y^i, i = 1, \ldots, 8e \log^{3/2}(n)/\log\log n$, *be as in Lemma 2.*

With probability at least $1 - o(1)$ *we have that for all* $c \in [\sqrt{\log n}]$ *there exists only one non-encoding* $j := j_{\ell+c} \in [n]\backslash\mathcal{B}(x, y)$ *with* $y'_j = 1 - y^i_j$ *for all* $i \in [8e \log^{3/2}(n)/\log\log n]$ *with* $\mathrm{Lo}_{z,\sigma}(y^i) = \ell + c - 1$. *Clearly,* $j = \sigma(\ell + c)$.

We are now ready to prove Theorem 1. As mentioned above, the proof also shows that the statement remains correct if we consider the unbiased black-box model with arbitrary arity.

Proof (of Theorem 1). We need to show that there exists an algorithm which maximizes any (a priori unknown) function $\mathrm{Lo}_{z,\sigma} \in \mathrm{LEADINGONES}_n$ using, on average, $O(n \log(n)/\log\log n)$ queries.

For readability purposes, let us fix some function $f = \mathrm{Lo}_{z,\sigma} \in$ LeadingOnes$_n$ to be maximized by the algorithm.

First, let us give a rough idea of our algorithm, Algorithm 4. A detailed analysis can be found below.

The main idea is the following. We maximize f block-wise, where each block has a length of $\sqrt{\log n}$ bits. Due to the influence of the permutation σ on f, these bit positions are a priori unknown. Assume for the moment that we have an ℓ-encoding pair (x, y), where $\ell \in [0..n - \lceil \sqrt{\log n} \, \rceil]$. In the beginning we have $\ell = 0$ and $y = x \oplus (1, \ldots, 1)$, the bitwise complement of x. To find an $(\ell + \sqrt{\log n})$-encoding pair, we first create a string y' with objective value $f(y') \geq \ell + \sqrt{\log n}$. By Corollary 1, this requires on average $O(\log n)$ queries. Next, we need to identify the $\sqrt{\log n}$ $f(y)$-critical bit positions $\sigma(\ell + 1), \ldots, \sigma(\ell + \sqrt{\log n})$. To this end, we sample enough bit strings such that we can unambiguously identify these bit positions. As we shall see, this requires on average $O(\log^{3/2}(n)/\log\log n)$ queries. After identifying the critical bits, we update (x, y) to a $(\ell + \sqrt{\log n})$-encoding pair. Since we need to optimize $n/\sqrt{\log n}$ such blocks of size $\sqrt{\log n}$, the overall expected optimization time is $O(n \log(n)/\log\log n)$.

Let us now present a more detailed analysis. We start by querying two complementary bit strings x, y. By swapping x with y in case $f(y) \geq f(x)$, we ensure that $f(x) > f(y) = 0$. This gives us a 0-encoding pair.

Let an ℓ-encoding pair (x, y), for some fixed value $\ell \in [0..n - \lceil \sqrt{\log n} \, \rceil]$, be given. We show how from this we find an $(\ell + \sqrt{\log n})$-encoding pair in an expected number of $O(\log^{3/2}(n)/\log\log n)$ queries.

As mentioned above, we first find a bit string y' with objective value $f(y') \geq \ell + \sqrt{\log n}$. We do this by running Algorithm 3, the (x, y)-encoded $(1 + 1)$ EA with mutation probability $1/\sqrt{\log n}$ until we obtain such a bit string y'. By Corollary 1 this takes, on average, $O(\log n)$ queries.

Next we want to identify the $\sqrt{\log n}$ ℓ-critical bit positions $\sigma(\ell+1), \ldots, \sigma(\ell + \sqrt{\log n})$. To this end, we query in the i-th iteration of the second phase, a bit string y^i which has been created from y' by flipping each non-encoding bit $y'_j, j \in [n] \backslash \mathcal{B}(x, y)$ independently with probability $1/\sqrt{\log n}$. If $f(y^i) = \ell + c - 1$ for some $c \in [\sqrt{\log n}]$, we update $X_{\ell+c} \leftarrow X_{\ell+c} \cup \{y^i\}$, the set of all queries with objective value $\ell + c - 1$, and we compute $\mathcal{J}_{\ell+c} := \{j \in [n] \backslash \mathcal{B}(x, y) \mid \forall w \in X_{\ell+c} : w_j = 1 - y'_j\}$, the set of candidates for $\sigma(\ell+c)$. We do so until we find $|\mathcal{J}_{\ell+c}| = 1$ for all $c \in [\sqrt{\log n}]$. By Corollary 2 this takes, on average, at most $8e \log^{3/2}(n)/\log\log n$ queries.

Thus, all we need to do in the third step is to update (x, y) to an $(\ell + \sqrt{\log n})$-encoding pair by exploiting the information gathered in the second phase. For any $c \in [\sqrt{\log n}]$ let us denote the element in $\mathcal{J}_{\ell+c}$ by $j_{\ell+c}$. We go through the positions $\sigma(\ell + 1), \ldots, \sigma(\ell + \sqrt{\log n})$ one after the other and either we update $y \leftarrow y \oplus e^n_{j_{\ell+c}}$ (if $f(y) < f(x)$), and we update $x \leftarrow x \oplus e^n_{j_{\ell+c}}$ otherwise. It is easy to verify that after $\sqrt{\log n}$ such steps we have $f(x) \geq \ell + \sqrt{\log n}$ and $f(y) \geq \ell + \sqrt{\log n}$. It remains to swap $(x, y) \leftarrow (y, x)$ in case $f(y) > f(x)$ in order to obtain an $(\ell + \sqrt{\log n})$-encoding pair (x, y).

1 **Initialization:**
2   **for** $i = 1, \ldots, n$ **do** $X_i \leftarrow \emptyset$;
3   Sample $x \in \{0,1\}^n$ uniformly at random;
4   Query $f(x)$;
5   Set $y \leftarrow x \oplus (1, \ldots, 1)$;
6   Query $f(y)$;
7   **if** $f(y) \geq f(x)$ **then** $(x, y) \leftarrow (y, x)$;
8 **Optimization:**
9   **while** $|\mathcal{B}(x,y)| \leq \lfloor \frac{n}{\lceil \sqrt{\log n} \rceil} \rfloor \lceil \sqrt{\log n} \rceil$ **do**
10       $\ell \leftarrow |\mathcal{B}(x,y)|$;
11   Apply Algorithm 3 with input (x, y) and mutation probability $1/\sqrt{\log n}$
        until it outputs a bit string y' with $f(y') \geq \ell + \sqrt{\log n}$;
12   Initialize $i \leftarrow 1$;
13   **while** $\exists c \in [\sqrt{\log n}] : |\mathcal{J}_{\ell+c}| > 1$ **do**
14          $y^i \leftarrow \mathbf{random}(y', x, y, 1/\sqrt{\log n})$;
15   Query $f(y^i)$;
16   **if** $f(y^i) \in [\ell, \ldots, \ell + \sqrt{\log n} - 1]$ **then**
17   Update $X_{f(y^i)+1} \leftarrow X_{f(y^i)+1} \cup \{y^i\}$;
18   Update $\mathcal{J}_{f(y^i)}$;
19          $i \leftarrow i + 1$;
20   **for** $c = 1, \ldots, \sqrt{\log n}$ **do** $\mathbf{update}(x, y, y', X_{\ell+c})$;
21   **if** $f(y) > f(x)$ **then** $(x, y) \leftarrow (y, x)$;
22   Apply Algorithm 3 with input (x, y) and mutation probability $1/\sqrt{\log n}$ until
      it queries for the first time a string y' with $f(y') = n$;

Algorithm 4. A $*$-ary unbiased black-box algorithm for maximizing $f \in$ LeadingOnes$_n$

This shows how, given a ℓ-encoding pair (x, y), we find an $(\ell + \sqrt{\log n})$-encoding pair in $O(\log n) + O(\log^{3/2}(n)/\log \log n) + O(\sqrt{\log n}) = O(\log^{3/2}(n)/\log \log n)$ queries.

By definition of Algorithm 4, all bit positions in $\mathcal{B}(x,y)$ remain untouched in all further iterations of the algorithm. Thus, in total, we need to optimize $\lfloor \frac{n}{\lceil \sqrt{\log n} \rceil} \rfloor$ blocks of size $\lceil \sqrt{\log n} \rceil$ until we have a $(\lfloor \frac{n}{\lceil \sqrt{\log n} \rceil} \rfloor \lceil \sqrt{\log n} \rceil)$-encoding pair (x, y). For each block, the expected number of queries needed to fix the corresponding bit positions is $O(\log^{3/2}(n)/\log \log n)$. By linearity of expectation this yields a total expected optimization time of $O(n/\sqrt{\log n}) O(\log^{3/2}(n)/\log \log n) = O(n \log(n)/\log \log n)$ for optimizing the first $k := \lfloor \frac{n}{\lceil \sqrt{\log n} \rceil} \rfloor \lceil \sqrt{\log n} \rceil$ bit positions $\sigma(1), \ldots, \sigma(k)$.

The remaining $n - k \leq \lfloor \sqrt{\log n} \rfloor$ bit positions can be found by Algorithm 3 in an expected number of $O(\log n)$ queries (Corollary 1). This does not change the asymptotic number of queries needed to identify z.

1 **Input:** An ℓ-encoding pair (x, y), a bit string y' with $f(y') \geq \ell + \sqrt{\log n}$, and, a
   set $X_{\ell+c}$ of samples w with $f(w) = \ell + c - 1$ such that
   $|\mathcal{J}_{\ell+c}| = |\{j \in [n] \backslash \mathcal{B}(x, y) \mid \forall w \in X_c : w_j = 1 - y'_j\}| = 1$;
2 **if** $f(y) \leq f(x)$ **then**
3       $y \leftarrow y \oplus e^n_{\mathcal{J}(\ell+c)}$;
4   Query $f(y)$;
5 **else**
6       $x \leftarrow x \oplus e^n_{\mathcal{J}(\ell+c)}$;
7   Query $f(x)$;

Algorithm 5. Subroutine $\texttt{update}(x, y, y', X_{\ell+c})$

Putting everything together, we have shown that Algorithm 4 optimizes any function $\mathrm{LO}_{z,\sigma} \in \textsc{LeadingOnes}_n$ in an expected number of $O(n \log(n)/\log \log n)$ queries. It is not difficult to verify that all variation operators are unbiased. We omit the details. $\square$

4   The Unbiased Black-Box Complexity of LeadingOnes$_n$

Next we show how a slight modification of Algorithm 4 yields a 3-ary unbiased black-box algorithm with the same asymptotic expected optimization time.

Theorem 2. *The 3-ary unbiased black-box complexity of* $\textsc{LeadingOnes}_n$ *is* $O(n \log(n)/\log \log n)$.

Proof. Key for this result is the fact that, instead of storing for any $c \in [\sqrt{\log n}]$ the whole query history $X_{\ell+c}$, we need to store only one additional bit string $x^{\ell+c}$ to keep all the information needed to determine $\sigma(\ell + c)$.

Algorithm 6 gives the full algorithm. Here, the bit string $\texttt{update2}(w, y', x^{\ell+c})$ is defined via $\left(\texttt{update2}(w, y', x^{\ell+c})\right)_i = w_i$ if $i \in [n] \backslash \mathcal{B}(y', x^{\ell+c})$ and $\left(\texttt{update2}(w, y', x^{\ell+c})\right)_i = 1 - w_i$ for $i \in \mathcal{B}(y', x^{\ell+c})$.

Note that, throughout the run of the algorithm, the pair $(y', x^{\ell+c})$, or more precisely, the set $\mathcal{B}(y', x^{\ell+c})$ encodes which bit positions j are still possible to equal $\sigma(\ell + c)$. Expressing the latter in the notation used in the proof of Theorem 1, we have in any iteration of the first **while**-loop that for all $i \in [n]$ it holds $y'_i = x^{\ell+c}_i$ if and only if $i \in \mathcal{J}_{\ell+c}$. This can be seen as follows. In the beginning, we only know that $\sigma(\ell + c) \neq \mathcal{B}(x, y)$. Thus, we initialize $x^{\ell+c}_i \leftarrow 1 - y'_i$ if $i \in \mathcal{B}(x, y)$ and $x^{\ell+c}_i \leftarrow y'_i$ for $i \in [n] \backslash \mathcal{B}(x, y)$. In each iteration of the second **while**-loop, we update $x^{\ell+c}_i \leftarrow 1 - y'_i$ if $\sigma(\ell + c) = i$ can no longer hold, i.e., if we have sampled a bit string w with $f(w) = \ell + c - 1$ and $w_i = y'_i$.

It is easily verified that Algorithm 6 certifies Theorem 2. We omit a full proof in this extended abstract. $\square$

1 **Initialization:**
2   Sample $x \in \{0,1\}^n$ uniformly at random;
3   Query $f(x)$;
4   Set $y \leftarrow x \oplus (1, \ldots, 1)$;
5   Query $f(y)$;
6   if $f(y) \geq f(x)$ then $(x,y) \leftarrow (y,x)$;
7 **Optimization:**
8   while $|\mathcal{B}(x,y)| \leq \lfloor \frac{n}{\lceil \sqrt{\log n}\rceil} \rfloor \lceil \sqrt{\log n} \rceil$ do
9      $\ell \leftarrow |\mathcal{B}(x,y)|$;
10   Apply Algorithm 3 with input (x,y) and mutation probability $1/\sqrt{\log n}$ until it outputs a bit string y' with $f(y') \geq \ell + \sqrt{\log n}$;
11   for $c = 1, \ldots, \sqrt{\log n}$ do
12   for $i = 1, \ldots, n$ do
13   if $i \in \mathcal{B}(x,y)$ then $x_i^{\ell+c} \leftarrow 1 - y_i'$ else $x_i^{\ell+c} \leftarrow y_i'$;
14   while $\exists c \in [\sqrt{\log n}] : |\mathcal{B}(x^{\ell+c}, y')| > 1$ do
15         $w \leftarrow \texttt{random}(y', x, y, 1/\sqrt{\log n})$;
16   Query $f(w)$;
17   if $\exists c \in [\sqrt{\log n}] : f(w) = \ell + c - 1$ then
18   for $i = 1, \ldots, n$ do if $x_i^{\ell+c} = y_i' = w_i$ then $x_i^{\ell+c} \leftarrow 1 - y_i'$;
19   for $c = 1, \ldots, \sqrt{\log n}$ do
20   if $f(y) \leq f(x)$ then $\texttt{update2}(y, y', x^{\ell+c})$ else $\texttt{update2}(x, y', x^{\ell+c})$;
21   if $f(y) > f(x)$ then $(x,y) \leftarrow (y,x)$;
22   Apply Algorithm 3 with input (x,y) and mutation probability $1/\sqrt{\log n}$ until it queries for the first time a string y' with $f(y') = n$;

Algorithm 6. A 3-ary unbiased black-box algorithm for maximizing $f \in$ LeadingOnes$_n$

5 LeadingOnes$_n$ in the Ranking-Based Models

As discussed above, we introduced two ranking-based versions of the black-box complexity notion in [3]: the *unbiased ranking-based* and the *unrestricted ranking-based black-box complexity*. Instead of querying the absolute fitness values $f(x)$, in the ranking-based model, the algorithms may only query the ranking of y among all previously queried search points, cf. [3] for motivation and formal definitions. We briefly remark the following.

Theorem 3. *The 3-ary unbiased ranking-based black-box complexity of* LeadingOnes$_n$ *is* $O(n \log(n)/\log\log n)$.

This theorem immediately implies that the unrestricted ranking-based black-box complexity of LeadingOnes$_n$ is $O(n \log(n)/\log\log n)$ as well.

   Theorem 3 can be proven by combining the Algorithm 6 presented in the proof of Theorem 2 with a sampling strategy as used in Lemma 2. Although the latter is not optimal, it suffices to show that after sampling $O(\log^{3/2}(n)/\log\log n)$

such samples, we can identify the rankings of $f(\ell + 1), \ldots, f(\ell + \sqrt{\log n})$, with probability at least $1 - o(1)$. We do the sampling right after Line 10 of Algorithm 6. After having identified the rankings of $f(\ell + 1), \ldots, f(\ell + \sqrt{\log n})$, we can continue as in Algorithm 6.

6 Conclusions

We have shown that there exists a 3-ary unbiased black-box algorithm which optimizes any function $\text{Lo}_{z,\sigma} \in \text{LEADINGONES}_n$ in an expected number of $O(n \log(n)/\log \log n)$ queries. This establishes a new upper bound on the unrestricted and the 3-ary unbiased black-box complexity of LEADINGONES_n.

Our result raises several questions for future research. The obvious one is to close the gap between the currently best lower bound of $\Omega(n)$ (cf. [1]) and our upper bound of $O(n \log(n)/\log \log n)$. Currently, we cannot even prove an $\omega(n)$ lower bound. Secondly, it would also be interesting to know whether the gap between the 2-ary and the 3-ary unbiased black-box model is an artifact of our analysis or whether 3- and higher arity operators are truly more powerful than binary ones.

Acknowledgments. Carola Winzen is a recipient of the Google Europe Fellowship in Randomized Algorithms. This research is supported in part by this Google Fellowship.

References

1. Droste, S., Jansen, T., Wegener, I.: Upper and lower bounds for randomized search heuristics in black-box optimization. Theory of Computing Systems 39, 525–544 (2006)
2. Lehre, P.K., Witt, C.: Black-box search by unbiased variation. In: Proc. of Genetic and Evolutionary Computation Conference (GECCO 2010), pp. 1441–1448. ACM (2010)
3. Doerr, B., Winzen, C.: Towards a Complexity Theory of Randomized Search Heuristics: Ranking-Based Black-Box Complexity. In: Kulikov, A., Vereshchagin, N. (eds.) CSR 2011. LNCS, vol. 6651, pp. 15–28. Springer, Heidelberg (2011)
4. Doerr, B., Winzen, C.: Playing Mastermind with constant-size memory. In: Proc. of 29th International Symposium on Theoretical Aspects of Computer Science (STACS 2012), pp. 441–452. Schloss Dagstuhl - Leibniz-Zentrum fuer Informatik (2012)
5. Rudolph, G.: Convergence Properties of Evolutionary Algorithms. Kovac (1997)
6. Mühlenbein, H.: How genetic algorithms really work: Mutation and hillclimbing. In: Proc. of Parallel Problem Solving from Nature (PPSN II), pp. 15–26. Elsevier (1992)
7. Droste, S., Jansen, T., Wegener, I.: On the analysis of the (1+1) evolutionary algorithm. Theoretical Computer Science 276, 51–81 (2002)
8. Doerr, B., Johannsen, D., Kötzing, T., Lehre, P.K., Wagner, M., Winzen, C.: Faster black-box algorithms through higher arity operators. In: Proc. of Foundations of Genetic Algorithms (FOGA 2011), pp. 163–172. ACM (2011)
9. Auger, A., Doerr, B.: Theory of Randomized Search Heuristics. World Scientific (2011)

Imperialist Competitive Algorithm for Dynamic Optimization of Economic Dispatch in Power Systems

Robin Roche[1], Lhassane Idoumghar[2],
Benjamin Blunier[1], and Abdellatif Miraoui[1]

[1] Université de Technologie de Belfort-Montbéliard,
Laboratoire Systèmes et Transports, 90010 Belfort, France
`robin.roche@utbm.fr`
[2] Université de Haute-Alsace,
LMIA / INRIA Grand Est, 68093 Mulhouse, France
`lhassane.idoumghar@uha.fr`

Abstract. As energy costs are expected to keep rising in the coming years, mostly due to a growing worldwide demand, optimizing power generation is of crucial importance for utilities. Economic power dispatch is a tool commonly used by electric power plant operators to optimize the use of generation units. Optimization algorithms are at the center of such techniques and several different types of algorithms, such as genetic or particle swarm algorithms, have been proposed in the literature. This paper proposes the use of a new metaheuristic called imperialist competitive algorithm (ICA) for solving the economic dispatch problem. The algorithm performance is compared with the ones of other common algorithms. The accuracy and speed of the algorithm are especially studied. Results are obtained through several simulations on power plants and microgrids in which variable numbers of generators, storage units, loads and grid import/export lines are connected.

Keywords: metaheuristic, imperialist competitive algorithm, dynamic optimization, economic dispatch, microgrid.

1 Introduction

With fossil resources becoming harder and harder to extract and worldwide demand continuously increasing, energy costs are expected to rise significantly in the coming years. Therefore, optimizing the use of these resources is of crucial importance in order to minimize costs. Such considerations are essential to power plant and transmission grid operators, for which optimizing operating costs can result in significant savings and profit. Economic power dispatch enables such an optimization and aims at determining the most cost-efficient and reliable operation of power systems, such as power plants. This objective is achieved by optimally dispatching available generation resources to supply the load connected to the system.

J.-K. Hao et al. (Eds.): EA 2011, LNCS 7401, pp. 217–228, 2012.

Optimization algorithms play an important role in economic dispatch, as they are the central tool used to obtain the optimal dispatch. Over the years, numerous algorithms have been used but imperialist competitive algorithms, a new type of metaheuristic based on imperialistic competition, are still largely unexplored for this problem. Moreover, very few were utilized in a dynamic optimization context.

The following studies whether this algorithm could and should be more widely used for this application. It describes the economic dispatch optimization problem, the imperialist competitive algorithm we used, how we tested it on mathematical functions and compared it with other common algorithms and finally how it was tested on two power systems for solving the dynamic economic dispatch problem.

2   Economic Power Dispatch Problem

2.1   Economic Dispatch Concept

The US Energy Policy Act [10] defines economic dispatch (ED) as "the operation of generation facilities to produce energy at the lowest cost to reliably serve consumers, recognizing any operational limits of generation and transmission facilities." In short, ED aims at optimally dispatching power generation between available generation units to meet demand. Its objectives are usually to minimize fuel utilization and sometimes also greenhouse gases emissions, compared to what less efficient generation sources would result in. Finding the best trade-off between costs, environmental impact and reliability is thus the main challenge of ED. In order to reduce operation costs, ED has been used in various forms for years, especially by transmission system and power plant operators.

Two types of ED can be considered: economic dispatch for the current day, and for future days.

- The first one, sometimes referred to as load following, consists in dispatching power for the current day, by monitoring load, generation and power imports/exports, while ensuring a balance between load and supply. The stability of the frequency of the grid (50 or 60 Hz) is the consequence of this balance, and is required by most loads which use it as a reference.
- The second one corresponds to dispatch for the following day, or several days after. Performed by the generation group or an independent market operator, it mainly consists in scheduling generators for each hour of the next day's dispatch, based on load forecasts. The units to use are selected based on their characteristics, costs and maintenance requirements.

The following will focus on the first type of dispatch, for the current day. Generators, such as gas turbines or fuel cells, will not be scheduled but will be sent set points their own control system should take into account immediately, while ensuring the reliability of the system. However, algorithms similar to the ones described in the following can be used for scheduling and committing generators. It should also be mentioned that determining if an algorithm behaves well for

ED is a necessary task, as the consequences of a bad ED can have a significant impact on costs for the operator and the consumer and cause instabilities or even blackouts.

2.2   Objective, Constraints and Hypotheses

To achieve an optimal power dispatch from an economic point of view, an objective function needs to be defined. This function (1) corresponds to the total cost of generation, storage, and power imports and exports. In order to maintain the grid frequency stable and the system reliability, constraint (2), requiring a balance between generation and supply, must be met.

$$\text{Minimize}\quad c_{\text{tot}}(t) = \sum_{i=0}^{n_{\text{gen}}} c_{\text{gen}}(P_{\text{gen},i}(t)) + \sum_{i=0}^{n_{\text{s}}} c_{\text{s}}(P_{\text{s},i}(t)) + \sum_{i=0}^{n_{\text{g}}} c_{\text{g}}(P_{\text{g},i}(t)) \quad (1)$$

$$\text{Subject to}\quad P_{\text{imb}}(t) = \sum_{i=0}^{n_{\text{gen}}} P_{\text{gen},i}(t) + \sum_{i=0}^{n_{\text{s}}} P_{\text{s},i}(t) - \sum_{i=0}^{n_{\text{l}}} P_{\text{l},i}(t) - \sum_{i=0}^{n_{\text{g}}} P_{\text{g},i}(t) = 0 \quad (2)$$

where $c_{\text{tot}}(t)$ is the total generation cost, n_{gen}, n_{s}, n_{g} and n_{l} are respectively the number of generating units, of storage units, of grid import/export lines and of loads. Their respective costs c_X and power outputs P_X use the same indexes.

In the following, constraint (2) is taken into account by the algorithm through its objective function. The actual fitness or objective function $f(t)$ (3) is a combination of the total cost defined in (1) and of the imbalance between load and supply. A coefficient α is set according to the ratio between the magnitude of the power system (W, kW, MW, etc.) and the estimated costs to give more importance to keeping the imbalance as low as possible while minimizing costs.

$$\text{Minimize}\quad f(t) = c_{\text{tot}}(t) + \alpha \cdot P_{\text{imb}}(t) \quad (3)$$

Additional constraints are also to be respected, notably for the optimization bounds. These bounds reflect the characteristics and dynamics of the controlled units. For example, a generating unit i such as a gas turbine can only operate within its operating range (4), and its power output variation is limited by ramp-up and ramp-down limits R_{u} and R_{d} (5). In other words, the bounds at time $t + 1$ depend on the ones at t because of the dynamic of the units and are thus updated at every call of the algorithm by an expert system.

$$P_{i,\min} \le P_{i,t} \le P_{i,\max} \quad (4)$$

$$P_{i,t} - R_{i,\text{d}} \le P_{i,t+1} \le P_{i,t} + R_{i,\text{u}} \quad (5)$$

In addition to these objectives and constraints, the following simplifying assumptions are made, as the primary focus of this paper is the performance of the algorithm:

- Valve-point effects, reactive power, line losses and emissions (e.g., CO_2 or NO_x) are not considered.
- All voltage magnitudes are assumed to be nominal. Bus and node capacities are assumed to be sufficient.
- Scheduling, starting and stopping generators is not achieved by the algorithm.

2.3   Problem Characteristics

The ED problem, as we will treat it, can be considered as a dynamic, non-linear, constrained, optimal control problem. It is at first dynamic because the cost function and the constraints change or can change at every call of the algorithm. The controlled power system is indeed itself a dynamic system, which evolves over time. A gas or wind turbine can for example be stopped for maintenance. Due to this dynamic nature, the algorithm is run at a fixed frequency, which can be as low as a few seconds. To a certain extent, it can also be considered as an online optimization problem, as the optimal control is performed in almost real-time. This is enabled by the small scale of the considered systems.

The variables to optimize are control set points corresponding to powers that are defined in continuous spaces, themselves defined by the energy management policy of the operator and by an expert system based on models. Algorithms able to operate in real-valued search spaces are thus required, as well as algorithms which do not require precise information on the search space or on the objective function. The reason for this is that the algorithm should be able to adapt to as many systems as possible without requiring much effort. Stochastic or randomized search methods are ideal for such constraints.

2.4   Optimization Algorithms for This Problem

The ED problem is a classical optimization problem in power systems and has been solved using numerous optimization algorithms:

- Linear programming [6,5], for which the objective function and the constraints must be linear. As we do not want to have any hypothesis on the shape of the functions, those algorithms are not selected. However, they are commonly used in the industry due to their speed and relative simplicity, but with constraint relaxation methods (such as Lagrange multipliers) and cost functions approximated by quadratic functions.
- Combinatorial algorithms, such as dynamic programming [11], can include integer variables and are based on the idea of splitting the problem into subproblems. Although these algorithms can be used for scheduling sources, their characteristics do not fit the needs of the current problem.
- Metaheuristics, such as genetic algorithms [12], particle swarm optimization [3] or simulated annealing. This last category of algorithms makes few or no assumption about the problem (e.g., the objective function does not need to be differentiable) but do not guarantee an optimal solution is ever found

as they rely on random variables. As we do not know how the objective function looks like and evolves over time, possibly with discontinuities, these algorithms are selected.

Although the variety of these algorithms is quite large, the literature contains few examples of imperialist competitive algorithms applied to the solving of problems in power systems [2], and none for the ED problem specifically.

3   Imperialist Competitive Algorithm

The imperialist competitive algorithm (ICA) is a new evolutionary optimization approach introduced in 2007 by E. Atashpaz-Gargari [1]. It is inspired by the imperialistic competition processes of human societies. The algorithm can be seen as a social counterpart of genetic algorithms. Several of its steps are indeed similar: countries can undergo revolutions as chromosomes can mutate, for example.

This algorithm uses a precise terminology, in which a solution is called a country. There are two types of countries: imperialist countries, and colonies, which depend on these imperialists. An imperialist and its countries form a group of countries called empire.

Algorithm 1. Imperialist Competitive Algorithm

1: Initialize the countries and form the empires
2: **while** *the stop condition is not satisfied* **do**
3:     **for** all empires **do**
4:         Move the colonies toward the imperialist (assimilation)
5:         Make some colonies undergo a revolution
6:         **if** a colony is more powerful than the imperialist **then**
7:             The colony becomes the imperialist and vice versa (overthrow)
8:         **end if**
9:     **end for**
10:     **if** two empires are too close **then**
11:         Merge them (unification)
12:     **end if**
13:     Make imperialistic competition occur
14:     **if** there is an empire with no colonies **then**
15:         Eliminate this empire
16:     **end if**
17: **end while**

ICA works as illustrated in Algorithm 1, where the following processes are used:

- Initialization and empire formation: Like other evolutionary algorithms, ICA starts with an initial population of solutions called countries, of size N_{pop}. Among them, the N_{imp} best countries (the most powerful) are selected to be imperialists. The remaining N_{col} countries form the colonies of these imperialists. The n initial empires are formed by dividing the colonies among imperialists according to their normalized power P derived from their cost c (6).

$$P_n = \left| \frac{c_n - \max_i c_i}{\sum_{i=1}^{N_{imp}} (c_n - \max_i c_i)} \right| \tag{6}$$

The number of colonies $N_{col,n}$ attached to empire n is computed according to (7).

$$N_{col,n} = \text{round}\,(P_n \cdot N_{col}) \tag{7}$$

– Assimilation: Imperialist countries attract colonies to them using the assimilating policy illustrated in Fig. 1. To update its position x, each colony moves toward its imperialist by updating its position using (8).

$$x_{t+1} = x_t + \beta \cdot \gamma \cdot r \cdot d \tag{8}$$

where $\beta > 1$ causes the colonies to get closer to the imperialist, $\gamma < 1$ corresponds to an assimilation coefficient, r is a random number chosen from the uniform distribution $\mathcal{U}(-\theta, \theta)$, θ adjusts the deviation from the original direction and enables searching around the imperialist and d is the distance between the colony and the imperialist.

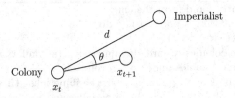

Fig. 1. Movement of a colony toward its imperialist

– Revolution: The revolution process introduces sudden random changes in the position of some countries. It plays the same role as the mutation operator in a genetic algorithm.
– Overthrow: After assimilation and revolution, a colony might reach a better position than the imperialist of the empire. In this case, the colony can become the imperialist and vice versa.
– Unification: If two empires are too close to each other, they can unite and become a single empire, with the sum of the colonies of the two initial empires.
– Imperialistic competition: Each empire tries to take possession of colonies of other empires and control them. This imperialistic competition is modeled by selecting the weakest colonies of the weakest empire and giving them to the empire that has the highest likelihood to possess them.

The total power $P_{tot,n}$ of each empire n is defined by the power of its imperialist plus its average colonies' power, as defined in (9) where $\zeta \ll 1$, I refers to the empire's imperialist and C to its colonies.

$$P_{tot,n} = P(I_n) + \zeta \cdot \text{mean}(P(C_n)) \tag{9}$$

The likelihood p_n, called possession probability, is then derived from each empire's power (10).

$$p_n = \left| \frac{c_{tot,norm,n}}{\sum_{i=1}^{N_{imp}} c_{tot,norm,i}} \right| \tag{10}$$

where $c_{tot,norm,n} = c_{tot,n} - \max_i c_{tot,i}$ is the total normalized cost of empire n and $c_{tot,n}$ its total cost.

In order to divide the colonies among empires based on their possession probability, a vector A is built (11), where P_n is the power of empire n and r_n a random value between 0 and 1. The selected colonies are then assigned to the empire whose relevant index in A is maximum.

$$A = [P_1 - r_1, P_2 - r_2, ..., P_{N_{imp}} - r_{N_{imp}}] \tag{11}$$

4   Performance on Mathematical Problems

Before testing the algorithm on the ED problem, its performance is compared to other evolutionary algorithms. These algorithms are:

- Metropolis Particle Swarm Optimization Algorithm with Mutation Operator (MPSOM) [4]: This hybrid PSO variant uses the Metropolis rule and a mutation operator to avoid local minima.
- Differential Evolution (DE) [8]: This algorithm combines the positions of solutions, called agents, to move them in the search space. Only the moves that lead to an improvement are accepted, others are discarded.
- Imperialist Competitive Algorithm (ICA), which was just presented.

The parameters of the ICA were empirically determined by running iterative trials using the mathematical functions described in Table 2, and starting with the parameters given by the authors in [1]. The tuned parameters are summarized in Table 1.

Several benchmark functions [9] described in Table 2 were used to test the algorithms performance. As the focus in primarily on the ED problem, the number of functions is limited to four. These functions provide a good start for testing the credibility of an optimization algorithm. Each of these functions has many local optima in its solution space. The amount of local optima increases with their dimension, which was set to 20 as in [7]. For each algorithm, the maximum number of function evaluations is $300,000$. A total of 30 runs for each algorithm was conducted and the average fitnesses of the best solutions were recorded.

Table 1. Parameter settings for the ICA approach

Parameter	Variable	Value
Number of initial solutions	N_{pop}	60
Number of initial imperialists	N_{imp}	6
Maximum number of iterations	N_d	5,000
Revolution rate	R_r	0.1
Assimilation coefficient	β	2
Assimilation angle coefficient	θ	$\frac{\pi}{6}$
ζ coefficient	ζ	0.02
Uniting threshold	U_t	0.02

Table 2. Standard benchmark functions adopted in this work

Function	Problem	Range
Sphere	$\sum_{i=1}^{n} x_i^2$	[-100;100]
Rastrigin	$\sum_{i=1}^{n}(x_i^2 - 10\cos(2\pi x_i) + 10)$	[-5.12;5.12]
Rosenbrock	$\sum_{i=1}^{n-1}(100(x_{i+1} - x_i^2)^2 + (x_i - 1)^2)$	[-2.048;2.048]
Ackley	$20 + e - 20\,e^{-0.2\,(\frac{1}{n}\sum_{i=1}^{n} x_i^2)^{\frac{1}{2}}} - e^{\frac{1}{n}\sum_{i=1}^{n}\cos(2\pi x_i)}$	$[-30.0; 30.0]$

Table 3. Comparison of the solutions obtained by the four selected metaheuritics. No lower threshold limit is set for the results.

Function	MPSOM	DE	ICA
Sphere	1.17×10^{-114}	7.145×10^{-40}	8.458×10^{-12}
	$\pm 1.17 \times 10^{-114}$	$\pm 1.390 \times 10^{-39}$	$\pm 2.532 \times 10^{-11}$
Rastrigin	0	4.551	2.251×10^{-06}
	± 0	± 0.919	$\pm 3.957 \times 10^{-06}$
Rosenbrock	1.44×10^{-02}	1.880×10^{-11}	0.3209
	$\pm 1.63 \times 10^{-02}$	$\pm 1.211 \times 10^{-11}$	± 0.4014
Ackley	1.16×10^{-10}	3.946×10^{-15}	8.025×10^{-16}
	$\pm 2.42 \times 10^{-11}$	$\pm 1.119 \times 10^{-16}$	$\pm 8.365 \times 10^{-15}$

The mean solutions and the corresponding standard deviations obtained for the algorithms are listed in Table 3. MPSOM obtains the best results for the first two functions, and DE for the third, Rosenbrock. ICA performs best on Ackley, with slightly better results than DE. A finer tuning of ICA parameters and hybridizing it with another algorithm would probably strongly improve its results on such functions. The following examines how the algorithm performs on the ED problem.

5 Simulations for the Economic Dispatch Problem

Two different grid configurations are used to test the performance of the algorithm, one is a microgrid where the stability of the algorithm will be tested and the other a fuel cell power plant where the optimization will focus on the total cost of the simulation. Both tests run over a period of four to five days and are based on real load profiles[1]. The optimization process is run every 60 s. Coefficient α from (3) is set to 10^5. The tests were run several times and returned very similar results.

5.1 Microgrid Test

The first test corresponds to a microgrid with two identical 83 kW fuel cells, photovoltaic panels (PV) with a rated peak power of 600 kW, two 500 kW wind turbines, a 1 MW import/export power line and a group of loads corresponding to the consumption of a residential area. The proportion of renewable energy sources in the microgrid corresponds to a very high penetration rate, which often implies control difficulties due to their intermittency. Simple models are used to extract the sources' power output from input values such as solar irradiation and wind speed[2], as well as to extract the hydrogen consumption of the fuel cells.

The expert system, which reflects the merit order chosen by the operator, determines the optimization bounds for each element. Renewable and non-controllable energy sources are used in priority and controllable generation units such as fuel cells are only allowed to provide the missing power to meet demand. Importing or exporting energy to the main grid is only allowed when the other sources are not sufficient (a similar rule could be used for storage).

The cost functions of the fuel cells reflect their fuel consumption, and generation costs from renewable energy sources are considered as equal to zero. The cost for importing a given amount of energy from the grid is lower than the price at which the same amount is bought by the grid, according to the principles of feed-in tariffs.

The test is run with three algorithms: ICA, DE and MPSOM. Results are shown in Fig. 2 for ICA and in Table 4 for all algorithms. They show that the algorithms perform well regarding the accuracy of constraint verification, i.e., the balance between load and supply (not displayed in Fig. 2 for the sake of clarity) is maintained. DE is the fastest but also the least efficient for minimizing total costs. MPSOM performs well, as well as ICA, except for its duration.

5.2 Fuel Cell Power Plant Test

The second configuration is a power plant based on a fuel cell array. A group of six 83 kW fuel cells generate power and supply a local load assimilable to an industrial and commercial area.

---

[1] Adapted from Southern California Edison's load profiles at:
  http://www.sce.com/AboutSCE/Regulatory/loadprofiles
[2] Extracted from: http://www.unige.ch/cuepe/html/meteo/donnees-csv.php

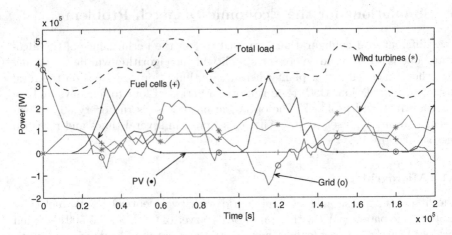

Fig. 2. Results of the microgrid simulation with ICA. Although intermittent sources induce large variations in generation, the use of fuel cells and of grid imports/exports, and the corresponding costs, are minimized by the algorithm while verifying the power balance constraint.

Table 4. Microgrid test results comparison

Algorithm	Unit	MPSOM	DE	ICA
Total cost	€	1,075	1,155	1,088
Mean imbalance	W	-0.0175	0.1344	0.1314
Mean duration	ms	8.126	4.075	257.0

The cost functions of the fuel cells differ from one fuel cell to the other. The aging and wearing of the units is taken into account by adding a multiplier coefficient η (12), which degrades or improves the cost of the fuel cells according to their age and how they are used and maintained. Coefficients between 0.9 and 1.2 are selected (see Fig. 3's legend – a high coefficient corresponds to an old and more expensive fuel cell), and are used to determine which algorithm behaves best for this kind of optimization.

$$c_{\text{fc}}(P_{\text{fc}}) = \eta \cdot m((P_{\text{fc}})) \tag{12}$$

where c_{fc} is the cost of the fuel cell for set point P_{fc}, and m the fuel cost function.

Table 5 summarizes the results of the tests shown in Fig. 3 for ICA. The same test profile is run with the three algorithms used in the previous test and a reference algorithm called equal dispatch (EqD), which simply consists in dividing the total load by the number of operating fuel cells without any optimization process.

Results show that ICA gives the best results in terms of total cost, as opposed to EqD. However, as with the previous test, it is also slower than the other algorithms. DE is particularly fast but is also rather imprecise, although this

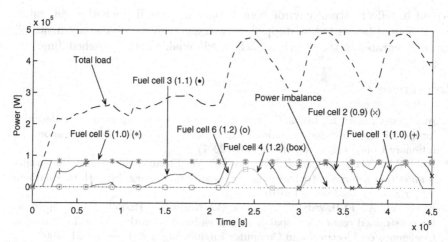

Fig. 3. Results of the power plant simulation with ICA. By properly selecting how much to use each fuel cell, the algorithm manages to achieve the best total cost. Older and less efficient fuel cells indeed tend to be less used by the algorithm to meet demand.

Table 5. Fuel cell power plant test results comparison

Algorithm	Unit	EqD	MPSOM	DE	ICA
Total cost	€	4,097	2,165	2,160	2,119
Mean imbalance	W	$\simeq 0$	-0.099	-2.286	-2.888
Mean duration	ms	< 1	11.82	8.391	454.8

imbalance only represents $10^{-4}\%$ of the maximum total load. It should also be noted that the imbalance for ICA is almost always close to zero except for a few minutes around 4×10^5 s, where the imbalance reaches 500 W.

6   Conclusion and Future Work

An application to dynamic ED of a new metaheuristic based on imperialist competition, the ICA, was described in this article. It showed that ICA provides good results for the ED problem, which was the aim of this work. It helps minimize costs even more than the other tested algorithms, while maintaining a good accuracy regarding the constraint. However, it is slower and performs less well on mathematical functions, illustrating the famous "no free lunch theorem". It therefore needs to be improved by further tuning its parameters and by hybridizing it with another optimization algorithm, similarly to what was achieved with MPSOM. Future work will also focus on speeding up the algorithm.

As the objective of this article was to focus on the performance of the algorithm for the dynamic ED problem, several simplifying assumptions were made. However, in the future, emissions of classical energy sources will be taken into account to reflect current environmental concerns. Deciding whether generation sources should be started or stopped, as well as demand-side management, will also be integrated in the algorithm, which will enable next-day scheduling.

References

1. Atashpaz-Gargari, E., Lucas, C.: Imperialist competitive algorithm: An algorithm for optimization inspired by imperialistic competition. In: IEEE Congress on Evolutionary Computation, pp. 4661–4667 (2007)
2. Duki, E.A., Mansoorkhani, H.R.A., Soroudi, A., Ehsan, M.: A discrete imperialist competition algorithm for transmission expansion planning. In: 25th International Power System Conference (2010)
3. El-Gallad, A., El-Hawary, M., Sallam, A., Kalas, A.: Particle swarm optimizer for constrained economic dispatch with prohibited operating zones. In: Canadian Conference on Electrical and Computer Engineering, vol. 1, pp. 78–81 (2002)
4. Idoumghar, L., Idrissi-Aouad, M., Melkemi, M., Schott, R.: Metropolis particle swarm optimization algorithm with mutation operator for global optimization problems. In: 22nd IEEE International Conference on Tools with Artificial Intelligence (ICTAI), vol. 1, pp. 35–42 (October 2010)
5. Irving, M., Sterling, M.: Economic dispatch of active power with constraint relaxation. IEE Proceedings C Generation, Transmission and Distribution 130(4), 172–177 (1983)
6. Nabona, N., Freris, L.: Optimisation of economic dispatch through quadratic and linear programming. Proceedings of the Institution of Electrical Engineers 120(5), 574–580 (1973)
7. Pant, M., Thangaraj, R., Abraham, A.: Particle swarm based meta-heuristics for function optimization and engineering applications. In: 7th Conf. Computer Information Systems and Industrial Management Applications, vol. 7, pp. 84–90. IEEE Computer Society (2008)
8. Storn, R., Price, K.: Differential evolution – A simple and efficient heuristic for global optimization over continuous spaces. Journal of Global Optimization 11(4), 341–359 (1997)
9. Suganthan, P.N., Hansen, N., Liang, J.J., Deb, K., Chen, Y.P., Auger, A., Tiwari, S.: Problem definitions and evaluation criteria for the CEC 2005 special session on real-parameter optimization. Tech. Rep. 2005005, Nanyang Technological University, Singapore and IIT Kanpur, India (2005)
10. United States Department of Energy: Economic dispatch of electric generation capacity (2007), a report to Congress and the States pursuant to sections 1234 and 1832 of the energy policy act of 2005
11. Waight, J., Albuyeh, F., Bose, A.: Scheduling of generation and reserve margin using dynamic and linear programming. IEEE Transactions on Power Apparatus and Systems PAS-100(5), 2226–2230 (1981)
12. Walters, D., Sheble, G.: Genetic algorithm solution of economic dispatch with valve point loading. IEEE Transactions on Power Systems 8(3), 1325–1332 (1993)

Author Index